Nature and The Human Condition

Nature and The Human Condition

A Study of Cultural Evolution

Cy Keedwell

Order this book online at www.trafford.com
or email orders@trafford.com

Most Trafford titles are also available at major online book retailers.

Printed in the United States of America.

ISBN: 978-1-4251-6843-8 (sc)

Trafford rev. 04/16/2011

 www.trafford.com

North America & international
toll-free: 1 888 232 4444 (USA & Canada)
phone: 250 383 6864 ♦ fax: 812 355 4082

For my wife, with love.

CONTENTS

CONTENTS

CONTENTS

Nature and The Human Condition

The opinions expressed in this book are the author's own unless noted otherwise. Every reasonable effort has been made to use only reliable data from trusted sources. However the accuracy or completeness of this work is not guaranteed nor is the fitness for use of other opinions or references given and it should not be assumed that their use by others would not infringe on privately owned rights.

INTRODUCTION

The human view of existence is brightened by eternal hope. It is a condition where Nature controls the terms and we explore the possibilities. Memories provide us with the necessary courage and determination.

Today we live in an era of civilized decline caused by an accumulation of unsolved problems. There have been repeated failures to find equality for all people, and a long period of economic expansion only caused more separation between them. That same process is now depleting resources and altering the world's climatic zones.

Humanity has experienced infinitely worse and survived but this is our time of maximum activity when we are most susceptible to harm. It would be fair to ask whether we can even spare the time to consider our plight or to prepare for any new dangers ahead.

This book offers an explanation of how and why this condition came to be, and how the future might unfold. Several years of research were spent to provide the background for new data, which are constantly emerging. It was not possible to cite every reference with the attention it deserved but I've made special mention of the ones on which key issues depend.

It all begins with a brief history of humankind, and proceeds to describe lifestyles more suited to the future limitations of our World. For completeness, many figures and some technical explanations are included but the text is intended for easy reading and contains all the essentials.

C. K.

Glossary words are printed in bold type in the text.
To ease visual scanning, other words and titles may be
printed in bold type indicating their inclusion in the
contents list, reference list, or index.

PART ONE

BELIEFS AND KNOWLEDGE

PART ONE

PART ONE

1.1 Perspective

The instincts we need are inherited at birth. At an early age we 'know' there are things that are good and worth defending, like families and access to nourishment and safety. By maturity we've also accumulated an array of physical and intellectual skills to help us function and survive in our cultures, which quite suddenly demand most of our attention.

At that age, everything we do involves decision-making and responsibility but it is our beliefs that determine our lives. Whether new beliefs could ever suppress our original instincts is debatable since that would require constant training by a dominant culture. Our daily beliefs are too transient to have such an effect, but generations of belief in organized religions has altered humanity's behavior and its cultures to some extent.

Our memories make us the most organized of all creatures. Anything causing doubt attracts our attention. Understanding usually involves checking the origin of a 'new idea', comparing it with our own experience, and arriving at an opinion by finding the logic and cause of it. I daresay most creatures are alike insofar as they cannot or will not live with doubt for very long. Instead of harboring uncertainty they are more at ease rejecting the new and confirming the old. In this way instincts remain intact, protecting the important sense of cultural-belonging.

The process of gathering these perceptions was never guaranteed to bring agreement between individuals or groups but it has provided a basis for debate and enabled human cultures to develop, in fact, almost all of our knowledge comes from others. For example, the philosophies of the western world are thought to have come from Hinduism, eastern Taoism, and ancient Greek democracy, but even those were the results of other smaller communities amalgamating and sharing beliefs. It seems the mixture of concepts helped the West to develop a global outlook several hundred years before it evolved in East Asia. We should also note that a reverence for Nature continued in the East while it declined in the turbulent development of the West. That is not to say that new beliefs are required for intellectual development. On the contrary, life can be better when naïve and not complicated by ambitions.

1.2 Spiritualism

One of humanity's distinctions is its accumulation of knowledge about the supposed laws of 'cáuse and effect'. An effect can have many causes, and one cause can have many effects. The chains of events seem endless and almost random, yet it does seem that life should be easy to understand. We certainly try to make it so by favoring the odds that it will proceed as we wish. That's how it was when humanity first started to think about such things.

The idea that life depends on invisible forces, started very early in human development. There were many versions, the most enduring being that spiritual entities preceded the world and after creation expressed themselves through it. In due course it was noted that human choices had their causes and effects, and they were remembered in belief systems that gave us shared purposes. Eventually they became organized religions: declared systems for combining spirituality with secular life, guidance other than that found in secular laws.

Whenever we have doubts it is good to remember that they are the price we pay for rapid cultural evolution. The older traditional beliefs of our cultures can be a timely comfort when we lack any other assurance. Sometimes the answers we need cannot be obtained from science, only through faith. Its guidance was enough for our ancestors and should be sufficient for us. Even so we have to accept that we survive not alone but in the midst of life in the World around us. That vital awareness can also be a comfort since it reminds us that beyond the risks of daily life, there are permanent values which have always been good. Living for those values brings us full-circle, back to the belief that all things have a purpose in Nature.

Figure 1 shows that about 75% of all people claim to follow one or other of the great religions. 10% believe in folk lore, the remaining 15% are agnostic or atheistic. Civilization's growth seems to have caused a reduction in the number of religions but, like the regions they serve, they've grown with their populations.

1.3 Civilization

Figure 2. The final sophistication was the decision of the majority to live in large collectives instead of familial tribes. Behavioral cultures grew out of the need for people to cooperate for safety reasons and to achieve shared goals. Collectives grew, each with a talent to retain, recall and use experience to answer their local needs. Eventually they established secular beliefs for routine endeavors. The merging of spiritual beliefs was not absolutely complete; there were schisms, lasting for

PART ONE

thousands of years. . . . But I'm getting ahead of the timeline:

The first recorded languages were in cave art about 50,000 years ago. Instead of words artists depicted people, hunting and searching for food or just gathered in groups. However these records are scattered and their time-lines are incomplete. There may have been 3,000 to 5,000 embryonic civilizations in world history. The number that actually survived into recorded history can only be estimated from the number of language roots still in use: about 250, with a total of almost 7000 adaptations to local needs.

Artifacts and oral history are certainly older than the first known civilizations such as the Yangshao (China), Harappa (Indus Valley) and Olmec (Mexico). Since human development was so gradual we find it useful to assume there was a point in time when civilizations found at least one purpose we can understand today. Though we could find even earlier starting points, in the West we often choose the **Agricultural Revolution** of 11,000YA. That would be within the last 2% of time since the emergence of *homo-sapiens*.

As agriculture increased, people migrated from savannahs, hillsides and coastal regions to live in river valleys and estuaries. From today's vantage point it seems logical that as populations grew, and as problem-solving became the main duty of leadership, the average per-capita **innovation quotient** would start to decrease. It did, and has continued to do so. Today it may be ready to *increase* as leaders are finding it more and more difficult to keep their influence without asking for democratic support. Even at 2500YA, discontented populations raised the question of which system was better: cooperating without question, or having the right

10

to voice one's opinion. Many centuries later, nations began to accept the idea of democracy. It still survives locally in its original common form, but has also been adapted for use by political factions in nation-states.

In towns, and later in homelands, layered systems of authority and trust became the normal way to implement the wishes of the majority of people. Funding had to survive an agenda of national projects before it could be spent locally. In this regard, referenda have gained importance though support for leaders is rarely determined by plebiscite. During elections we depend on philosophies directed to resolve common needs, then new leaders have to rely on trust in their decisions rather than wait for popular votes on every detail.

1.4 World Economics

Yes, economics is a belief system! Before it was a recognized discipline it already included the belief that each person needed to work in order to survive. Everyone contributed, resting only to regain the strength to return to their tasks. Their reward was in sharing the benefits with those who depended on them.

Since the European Renaissance, economics has been consciously used by most cultures in the belief that it was a humanitarian necessity. The moral principles were as old as family life: help the needy and save for a rainy day. But in our more-international world it seems that that intention is defeated by the unequal distribution of benefits within and between nations.

In addition to these complications profit is always measured within specified activities, a fact which tends to limit its universality. For example, imagine the insular economy of a remote island: As its natural resources are depleted it operates at a loss, yet for some occupants 'profit' exists in the form of increased ownership of things others need. This concept wouldn't be a problem if everyone's efforts were equally reimbursed but a few always claim the right to work faster or invest more capital or negotiate more favorable terms of employment, all to secure a greater share of profits. Given these circumstances the island's economic future is likely to be determined by whoever accumulates ownership of whatever remains, not by whether any action helps or harms the populace.

However, the fact that there is always a net loss of resources shouldn't deprive anyone of employment at a fair rate of pay. There is always much to be done. The problem is in finding a way to share the work and its benefits in an equitable way. Cash payment is not a reliable measure as one worker's pittance is another's fortune. The more important metric is the time spent working toward shared goals.

On our hypothetical island the net profit might then be expressed in terms of collective progress, and *each group would be reimbursed for the per-capita time it contributed.* It might be the same when nations trade among themselves; ideally they would be paid in proportion to their populations, and by the per-capita time spent working toward world goals.

Actually, placing trust in nations to work toward any world goal has proven difficult as there is no obvious benefit in cooperation except for the possible reduction of environmental loss. That concern is expressed under several headings such as ecology, recycling, renewable energy, and environmentalism, the only differences being in how the human condition is perceived and how people intend to adapt. There are tasks that nations might share today but completing them and sharing the benefits will take more than a lifetime, which is why they are not enthusiastically supported.

Growth and The Main Economies

The flaw in modern civilization is that its desire to grow is used to justify whatever harm it causes, and the resultant acceleration causes loss with only a token effort being expended on conservation and replacement. The world's population is growing exponentially and trade is expected to redouble in less than 100 years. The data will show that this is less than **sustainable**, and may even be terminal in the not too distant future.

The world is at a crossroads, and each nation should be involved in reducing the environmental impact of its activities. It is better to be part of the natural world and save it than to prosper in isolation and watch it decline. As to the future of civilization, this era seems to be the only time when we shall have so many opportunities to work collectively for its well-being.

Thus far each culture has been able to develop its own economy: Some nations sell their resources without the safety net of a sustainable income-base and then, when irreplaceable reserves were gone, they developed 'hollow economies'. Others have reliable income bases but need to import materials to conserve their own reserves. In general the latter are more likely to keep their cultures intact.

The formation of alliances represents a stage in the formation of global organization which helps nations to preserve their identities. About half of world-trade is handled in this way. Most recently, Brazil, Russia, India and China, already members of other trade blocs, formed a policy alliance called **BRIC** to increase their global influence. They have about 40% of the world's population and 30% of its **GDP(PPP)**.

PART ONE

The following list shows the main economic details of five traditional geo-political regions:

EUROPE. The Industrial Revolution provided the money for new construction, foreign investments, social services, armed forces and futuristic planning. Colonial investment was attempted to expand European influence, until it was found to be too difficult; the colonies grew to rely on their own heritage for guidance and now manage their own affairs. The European Economic Community (**EEC**) was formed in 1957 and in 1992 was renamed the European Union. People with an **EU** passport can live and work in any member country without asking for permission. Unemployment rates are 5 to 10%.

ASIA has the worlds largest national populations. It also has the widest range of ethnicity, resources and human activities. The **GDP** per capita remains low. Urban unemployment is 2 to 10% and rural unemployment is 10 to 30%. Its three major powers are:

RUSSIA. The world's largest country spans the European and Asian land mass. The independence of former Soviet-Communist states was declared in 1991. Today, their individual economies vary: Industrial and richer in the West, rural and poorer in the East.

INDIA. Achieving independence from imperial rule in 1947, it continues to develop economically and hopes to regain equality with China in world trade. It has the world's second largest but fastest growing population, which has virtually filled its livable habitat, and is now conserving food.

CHINA. With the world's largest population, slightly more than India, China is growing its industrial sector. Its total economy is divided between modern ideas and traditional farming. It needs more coal, softwood lumber, fresh water, grain, fisheries and arable land, and the importation of resources makes it increasingly international. Its growth may even be too fast for its central government to control as it would wish.

China's air pollution and additions to **global warming** are substantial and increasing. Energy consumption is about one third of all other nations combined, six times that of the USA and three times that of India. Most of it comes from coal.

Population is 1.32 billion: about 20% of the world's total and 4.4 times that of the USA.

Agricultural land is 1.4 million square kilometers: less than 7% of the World's total.

National **GDP(PPP)** is presently about 10.2 trillion US dollar equivalent and 15.5% of the World's total.

NORTH AMERICA AND OCEANIA. The modernization of this entire region occurred at a time when the rest of the world was already nearing economic completion. Its **per-capita GDP(PPP)** is the highest of any geographic region. Foreign direct investments are attracted by the availability of natural resources. Immigrants are used to keep its economies growing in relation to the rest of the world. Today the main task is to stay competitive while replacing ageing infrastructure. Unemployment is in the range of 5 to 10%.

SOUTH AND CENTRAL AMERICA. Like North America, this region is rich resources and its modernization was helped by the arrival of immigrants who brought their own traditions with them. However it has the largest percentage of indigenous people outside of the Old World and this helped it to retain more of its pre-Columbian culture. Unemployment is usually in the range of 5 to 15%.

15

AFRICA. The long-term history of the continent seems to be one of foreign exploitation and economic decline. The African Union was formed in 2002 as the successor to the Organization of African Unity which began in the 1960's. Its forum is the Pan African Parliament with members from almost all nations on the continent. North African unemployment is typically 15 to 30%. Data from sub-Saharan Africa tend to be incomplete and the extent of its unemployment is hard to define; national estimates range from 25 to 85 % of their workforces.

1.5 International Organization

There is no virtue in the idea of a single world culture. That possibility has been tested at various times in various ways using methods such as assimilation, conversion and dominant occupation. The results were always the same: you cannot achieve peaceful accord by superimposing one culture on another.

At the other extreme isolationism is no better, as it can impose excessive uniformity *within* cultures. This error can be seen in the world of international commerce where trade takes precedence over humanitarian principles. In terms of real value there is always more loss than gain, which is compensated by distributing the burden onto the shoulders of the populace. Interestingly, this parallels the previously mentioned effect of the loss of resources in isolated economies; profits are shared among the principle-owners and there is little or no progress in human development among the rest.

Most countries accept that the benefits of international trade, alliances, political associations and defense agreements are worth such an 'inconvenience', but in order to have a share of the benefits they also need to develop certain standards of sophistication. For example, a modern capital city is the showpiece of many developing countries, followed by international airports, government buildings, transportation systems, law enforcement, armed forces and wireless communication networks. These infrastructures are costly to build, operate and maintain, so the result of economic development is often unequal human development (measured from basics such as nutrition, education, access to telephone service, clean water distribution, and even sanitation). The **UNDP** has issued Human Development Reports (**HDRs**) for the last 20 years. There have been improvements in the human development indexes (**HDIs**) of developing countries but little convergence of nations toward a world average. To resolve this problem, each nation should be able to assess its own value by comparing its *useful per-capita effort* to a world average. National entitlements should then be proportional. Greater coordination between countries on the sharing of real values would be a first step in solving many humanitarian problems. Many believe the world should be reorganized as a federation of cultures rather than a regulated, but only semi-compliant, amalgam. This will be discussed later.

The present aim is to achieve standards known as **Millennium Development Goals** (**Figure 3**) but the test period is brief (15-20 years).

The UN as a Belief System
Total world peace is rare. The process of achieving it involves bringing national representatives together to make their grievances known within an assembly of shared principles, The **UN Charter (1945)** and the Universal Declaration of **Human Rights (1948)** contain

17

the essentials, and success depends on the amount of international support they receive. As shown in **Figures 4, 5** and **6**, the **UN** provides a wide range of services and has an important place in the lives of all people. Peace keeping, financial order, international trade and law, are just four of its major priorities. It is appropriate to take the time here to review its development and understand its form and function in global affairs.

The Charter was adopted in 1945. It was needed to build an international democracy, organize monetary policy, manage post-war recovery and restore international diplomacy. The **IBRD** (later part of the **World Bank** with the **IMF**) was formed to provide financial consultancy and assistance.

The World Court reflects the majority of world opinions on international law though in practice it rarely challenges the sovereignty of nations. There is a **Security Council** which speaks to national governments and responds to complaints of human rights violations or armed conflicts. This Council has fifteen members. Of these China, France, the Russian Federation, the United Kingdom, and the United States of America are appointed with powers of veto. The others are elected for two year terms without the power of veto.

The General Assembly is the setting for World debate. This large system is manageable because representatives follow the known policies of their governments. There are many **UN** Commissions with international membership whose concerns include a wide range of humanitarian topics on behalf of all nations. There are also many Operating Divisions and sub-divisions which publish statistics. e.g. **UNEP, UNICEF, UNAIDS, ILO, FAO, UNESCO, WHO, WMO, WTO, DESA,** and the **World Bank** Group. Many have their own discretionary powers and with such a wide range of activities they have to maintain a careful balance.

International organizations such as **WEC, OECD, OXFAM, ICRC, Amnesty International** and **The Global Policy Forum** have their own means of publication and represent economists, trade groups, philanthropic groups, emergency response organizations, scientists and others. Public factions can offer their own opinions on issues which their governments might view differently. Some are formally recognized and may even have **UN** reporting privileges. For convenience they are called 'non-governmental organizations' **(NGOs)**. A Non-Governmental Liaison Service **(NGLS)** publishes 'Roundup', a series of papers on items of general interest.

Anyone with access to the internet can find websites for the UN commissions, divisions and related groups as well as those of numerous **NGOs** and 'think-tanks'. One can search using their acronyms or their main areas of interest. It is advisable to look for opposing views on any subject. Samples of public opinion are found in the media and social networks.

1.6 Conclusion of Part One

In a relatively short time-span our species built an artificial reality which can be perceived in a thousand different ways, because cultural beliefs vary and experiences influence our perceptions. Translating this into a common format for debate has always been one of the weaknesses of our diverse civilizations. If it works reasonably well it is because humanitarian principles enable us to bring all but the most extremist visions into focus. Without those principles it will deteriorate and ultimately fail.

PART ONE

Sharing experience is the key to better understanding. The time for our world to function as a single economic entity is not too distant to consider, perhaps a hundred years or so. The ultimate reason for this change will be that the present format does little for humanity, and with its competitive regime actually reduces the freedom of diverse cultures to live as they would wish.

The present fast rate of economic growth with its slower rate of human development actually separates people who could be cooperating. In the present absence of a world democracy of equals, national governments have the difficult task of working to secure some of the global benefits for their homelands. They can declare their intentions to support world initiatives, yet for economic reasons be unable to make much progress. We've mentioned possible improvements to the way benefits are distributed on 'Island Earth'.

BELIEFS AND KNOWLEDGE

Figure 1

COMMON NAME OR DESCRIPTION	APPROXIMATE TIME OF FIRST RECOGNITION (YEARS AGO)	FOLLOWERS AS % OF WORLD POPULATION IN 2000CE
NON-RELIGIOUS	---	12.0
ATHEISM	---	2.4
FOLK RELIGIONS	10000	6.0
HINDUISM	5000	13.0
JUDAISM	4000	0.25
TAOISM	2600	---
CONFUCIANISM	2550	0.1
BUDDHISM	2500	6.0
CHRISTIANITY	2000	33.0
ISLAM	1390	20.0
SHINTOISM	1290	0.1
SIKHISM	530	0.4
OTHERS	---	6.75

The Main Spiritual Beliefs

PART ONE

Figure 2

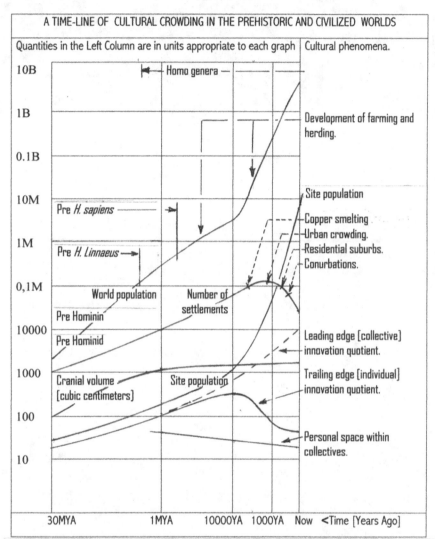

A TIME-LINE OF CULTURAL CROWDING IN THE PREHISTORIC AND CIVILIZED WORLDS

The Crowding of Civilizations

PART ONE

Figure 3

EIGHT GOALS briefly stated:

1. Eradicate extremes of poverty and hunger.

2. Achieve universal primary education.

3. Promote gender equality and empower women.

4. Reduce child mortality.

5. Improve maternal health.

6. Combat HIV-AIDS, malaria and other diseases.

7. Ensure environmental sustainability.

8. Organize a global partnership for development.

Note: 48 progress indicators are being monitored to achieve specific goals by 2015 (or 2020 in one case).

The UN Millennium Development Goals

PART ONE

Figure 4

SECURITY COUNCIL, GEN. ASSEMBLY, INTERN'TNL COURT OF JUSTICE, SUBSIDIARY BODIES, SECRETARIAT, TRUSTEES,
Related organizations: WTO(trade), IAEA, CTBTO, OPCW,
Specialized agencies: ILO, FAO, UNESCO, WHO.
World Bank Group: IBRD, IDA, IFC, MIGA, ICSID,
Other organizations: IMF, ICAO, IMO, ITU. UPU, WMO, WIPO, IFAD, UNIDO, WTO(tourism).
Depts. and offices: OSG, OIOS, OLA, DPA, DDA, DPKO, OCHA, DESA, DGACM, DPI, DM, OHRLLS, UNSECOORD, UNODC, UNOG, UNOV, UNON.
Other entities: OHCHR, UNOPS, UNU, UNSSC, UNAIDS.
Research and training: UNICRI, UNITAR, UNRISD, UNIDIR, INSTRAW.
Programs and funds: UNCTAD, UNDCP, UNEP, UNICEF, UNDP, UNFPA, UNHCR, WFP, UNRWA, UN-HABITAT.
Subsidiary bodies: military, monitoring, peacekeeping, inspection, crime, tribunals, compensation, standing & ad-hoc committees.
Commissions: human rights, crime, technology, population, development, statistics, economics, environment, etc.

Lists of acronyms are after the glossary at the back of the book

The United Nations [2004]

PART ONE

Figure 5

Date formed: 1945   Universal Declaration of Human Rights: 1948

Member nations...192
Non-member states...10
Non-member territories ...38
Non-member entities...5

Principle organs:

Security Council...Five permanent members with veto privileges.
...........Ten non-permanent members.

General Assembly...........................All member nations.

International Court of Justice...Fifteen judges.

Economic and Social Commissions

Secretariat

Trustees

Employees World-wide...65000 approximately.

Financial income.Estimated in 2007 to be 20 billion International Dollars at 2005 values. 7512 million came from member assessments and 12488 million came from additional contributions. This income was 0.033% of World GDP. About two thirds of it was paid by only ten nations, which produced 40% of the World's GDP with 11% of the World's population.

United Nations Statistics   (2007)

PART ONE

BELIEFS AND KNOWLEDGE

Figure 6

INFORMATION SOURCES. MEDIA AND PROFESSIONAL RESEARCH/ DELEGATIONS. Non-governmental organizations, NGOs with reporting privileges, Representative of member states.

THE UNITED NATIONS WORLDWIDE ORGANIZATION		
SECURITY COUNCIL.	**GENERAL ASSEMBLY**	
Intervention. Monitoring. Peacekeeping. Weapons control. Other issues.	Representatives of member states, assistants, advisors, secretarial services.	
	ECONOMIC AND SOCIAL COMMISSIONS.	**SECRETARIAT**
	Human rights, Health, Crime, Trade. Banking, Loans, Other issues.	
Factual research, data and opinion gathering, special funds, assistance, outreach, publications.		**INTERNATIONAL COURT OF JUSTICE**
		TRUSTEES

Non-governmental and multinational organizations
W O R L D   P O P U L A C E

The United Nations and World Interactions

PART ONE

PART TWO

NATURAL ADAPTATION

PART TWO

PART TWO

2.1 Perspective

The scientific view is that life began when Earth was cooling and filled with changing conditions. There must have been many beginnings and ends before life could be sustained. Perhaps it only occurred when there were enough microorganisms for some to survive change and recover from local losses. In great numbers they could have modified their own habitats chemically. The main evidence for this was a change in the atmosphere, from carbon dioxide-rich to oxygen-rich:

About 2000 MYA these gases were present at about 10% and 0.2% by volume respectively, and by about 700 MYA they were transposed to about 0.03% and 20%. For the last 170 million years the range for carbon dioxide has been 0.3 to 0.02%, and for oxygen 27 to 21%. Biologically, it began when small cells were taken into larger ones which survived in symbiosis. In time, the small cells were expressed as organelles in the reproductive process. Later specialization caused the flora-fauna divergence. Plants now obtain chemical energy from sunlight and emit more oxygen, and animals obtain chemical energy from plants and emit more carbon dioxide.

Such phenomena altered the world slowly, but now our own civilization has become destructive, rapidly depleting ocean life, scraping the land bare for building and agriculture, and accelerating polar ice-melt. We call this the **anthropogenic effect**, as if it were an unavoidable consequence of life. Indeed, until the last few years we were not even aware of the extent of civilization's effect on the natural world. Our only practice had been to exclude 'non-useful' species from our habitat, but now our population density is enough to keep all but parasitic species at a 'safe' distance; so-called 'wild animals' are comparatively rare visitors. If life seems to grow harsher it is the result of our own actions. We accomplished more than we could have imagined and now live with the consequences.

Life itself is not always kind: Within the last 2000 years, entire cultures disappeared and fertile lands changed into deserts or were washed away by floods, all without human involvement. Several times per year we learn about new physical disasters or epidemics for which human activity was not the initial cause, yet there have been other events which were accepted almost without question, simply because we did cause them.

2.2 The Process of Adaptation

With or without human involvement, **climate change** is now the most enduring force affecting life on Earth. It can proceed slowly, for example in millions or thousands of years due to continental drift, altered ocean currents and variations in Earth's solar orbits (**Milankovitch** cycles), or quickly, in less than one hundred years, when it can be linked to one or more catastrophic events altering Earth's reflectance of solar radiation in the atmosphere and at ground level (its **albedo**). Typical causes are large volcanic eruptions and **bolides**. Given enough time such events are followed by natural cycles of change as the World tries to adapt. Some species thrive, others are diminished.

NATURAL ADAPTATION

The emergence of sub-species is often attributed to an imaginary **molecular clock,** but genetic mutations are available at all times. Sub-species only survive when there is a niche for them to fill, usually after the extinction of a member of its own **clade.** Therefore it seems to me that the process is more dependent on the failure of incumbent species than on challenges from mutants. However, when this adaptive selection occurs it happens fastest in species with short life cycles, such as bacteria. In creatures of the higher species with life cycles measured in decades, biological change can take millions, or at least thousands of years. Entire species may even disappear before they can adapt to changing conditions.

We find ourselves in a relatively moderate but similar situation. The present rate of climate change is trending beyond our limited experience. Our species has an innate ability to adapt behaviorally with a minimum of effort, but the World has its own cycles of change and given enough time, we shall have to comply biologically.

Human activity has been increasing exponentially for more than 10,000 years and has accelerated dramatically in the last 200, but in that time we have contributed far less to Earth than we took. Our affect on the **biome** is clearly visible and is a catastrophe in-the-making. We survived ice-ages and grew in warmer climates but have yet to show signs of slowing down, which at least would make life easier and minimize natural repercussions. It seems these are lessons we prefer to learn by trial and error, though we are certainly capable of learning them formally. Is it Earth's function to support us, or is it the other way around? Biologically there is no starting point and no foreseeable end; if any species fails to contribute, it doesn't survive and others take its place.

PART TWO

2.3 Human Experience

The climate was at its coldest about 130,000YA, then a relatively fast thaw allowed migrations into central Asia and Europe. This was followed by progressive cooling and more ice ages.

Between 60,000 and 30,000YA, the average sea level was at least 100 meters lower than it is today. Land bridges and sea ice connected many islands to their continents and much of the human population was enabled to live on dry continental shelves; only rivers and mountain ranges interfered with communication. Traces of that era have been gathered from inland sites. Some tribes found places to settle, others followed the migratory routes of large animals, occasionally arriving in the Americas (North and South). Similarities between Asian and early American cultures show the extent of this activity. The wonder is that humanity took so long to explore the western half of the world.

Starting about 18,000YA continental ice sheets melted, raising the sea level 130 meters, flooding habitats and submerging possibly the most interesting part of world history: the flooding of the Mediterranean Basin was just such an event. In time, mountain snow lines rose thousands of meters opening new habitats. In the high plains of the Andes, fresh water supplies increased due to melting glaciers then stopped. Glacier-fed lakes vanished or evaporated leaving salty residues.

In other parts of the world, river courses became seasonal streams or permanently dry and regional cultures disappeared as tribes could no longer stay where their ancestors had settled. Irrigation-farming became a common type of work. Melanesia was perhaps the first region to practice wetland cultivation. The rising sea level must have been slow and hardly detectable, but each generation had to adapt to new limitations. Their dispersal to other islands in Oceania became an accepted way of life. Sea-going explorations and migrations became traditional in their cultures.

The format of what is now called modern Western culture evolved in Southern Asia was modified in Mesopotamia and then spread into Europe. During this time, settlements grew along trade routes and were held by occupational force. However peace was not assured as towns-people had to protect themselves from marauders. Life began to improve only when local authorities were able to enforce law and order. A few cultures were unable or unwilling to find peace and continued with piracy and rebellion.

The re-discovery of North America by a second wave of migrants started only in the last 1000 years and still brings people from the rest of the world. The freedom to live and work in whichever region suits us best may be as close as we shall ever get to complete liberty for the individual, however we are still centuries away from achieving it as a human right. Growing populations always had to search for more food and fresh water and that search continues today. Our rates of migration are small compared to what they might be in future. This will be discussed later in the book.

2.4 Economics

Economists study average gains and losses for any number of purposes. Uniformity of distribution is not a prerequisite but there is morality in trying to achieve it for all people. Humanitarianism calls for the majority to share most of the benefits of a collective effort.

The terms 'cash-flow' and 'wealth' are often confused. Wealth is a potential linked to ownership; it is static and requires proof of value. Cash-flow is the rate at which ownership is exchanged; it is dynamic and usually valued as the reciprocal of its rate of circulation. The average income of a collective depends on its cash-flow, or more realistically, the deprivation of its majority increases with the absence of cash- flow.

One way to compare the economies of nations is to look at the percentages of **GNP** spent by their governments on social services. **Figure 7** shows three stages of national development: Complete, Intermediate and Beginning. The disparity ratio is 15 to 1 for GNP spending and 40 to 1 for health care alone. Another way is to compare national per-capita earnings. **Figure 8** shows that the average earnings, as **GDP (PPP)** per capita in developed nations, is 6 or 7 times higher than in less developed ones, but international comparisons hide even greater inequalities. For example in the USA and many other countries there are large groups with personal incomes above one million dollars per year, and others with 3000 dollars or less. Similar disparities exist in nearly all countries. The poorest, or most heavily indebted, have large groups with no homes and no income.

The less extreme way to view economic disparity is simply to consider the variance between nations without reference to conditions *within* their populations. The national averages of per-capita income plus-social benefits are then seen to vary by a ratio of at least 20-to-1. Even so, could the world be any further from economic equality? It seems there is a stage of economic growth beyond which disparity causes less concern. Leading groups learn how to increase their earnings faster but it is an advantage they hardly ever relinquish. A brief history of economic development illustrates this point as a cause, followed by an effect:

NATURAL ADAPTATION

Three hundred years ago, as the world's rate of **GDP** growth increased, agrarian cultures began to decline (**Figure 9**). This was through no fault of their own: Western economies became the temporary leaders and the East-West gap widened as Europe and the New World added new technologies. Eventually, the work of European and North American farmers was mechanized and technology opened new employment opportunities in cities. Factory employment took the agricultural labor overflow and the industrialization of the West was soon completed. Reconciliation between the world's industrial and agrarian nations was not attempted until much later, after two world wars had occurred.

1945 was the year in which the **UN** was formed. The economy had slumped and more international trade was needed. It was the era of **Breton Woods** planning, designed to increase trade by offering loans to developing countries. Most of them derived financial benefits but heavily indebted nations began to fail because their skills were unsuited to the task of rapid recovery. Certainly there were resources and commodities which were in great demand, yet for some, loan repayment involved putting workers into the factories of foreign corporations.

The result of this activity was the rapid spread of western consumerism around the world accompanied by a substantial loss of cultural diversity. We cannot be sure that less funding would have made developing countries more resilient than they are today. If there is blame to be placed it is on the events which caused the wars and changed the world.

Prior to 1960 there was no reason to doubt the correctness of our plan to revive economic growth; we were generally unaware that environmental problems could hurt us. But now the real condition has become obvious: We moved too quickly and Nature responded with more global warming, leaving us with the task of adapting to the conditions we caused yet without knowing how that effort might again alter them.

2.5 Sharing Benefits and Burdens

Countries will always have their own points of view, but cultures should be able to learn from the experience of others without necessarily looking for a competitive advantage. Proprietary knowledge is a case in point: It can be either shared or misused. Perhaps it is only in humanitarian disciplines such as health care, that knowledge is treated as a global resource. Every effort should be made to share all resources which improve the human condition.

No nation or consortium should force its own ideas onto another. All people are team players in human development, whether or not they contribute to its success. In regard to the world's constant desire for financial growth, this is a good time to ask whether the disadvantages outweigh the benefits:

Unequal development, especially in productivity, separates nations but ergonomic efficiency is also a factor. In finance the three basics: energy, material and labor are only convenient divisions, and then only when there is enough work for everyone. The international solution is to help poorer national economies to grow by providing technology, ensuring that there is an abundance of energy in the world, and buying products from countries which are under-employed. Increased efficiency and low-paid workers might provide a competitive edge, at least temporarily, but the moral purpose of economic growth is not to increase **GDP** as such. Instead it should build a standard of global equality and ensure that each nation's cash-flow grows with its population. Failure to do so causes greater delays in economic recovery, increases disparity, and causes other negative effects such as the marketing of low-quality products, wasted resources and above all, misdirected human effort.

2.6 World Limitations

Worst-case scenarios for the human population were foreseen long ago, and have been revised at intervals throughout history. Perhaps the most frequent references concern the food supply, which will limit growth. However, before that time arrives it will be normal to employ more agricultural workers wherever there is the potential to increase the world's food output. Clearly, economic factors and climate change will be involved.

Figure 10 shows estimates of regional populations from year 1800 to 2100 without any purposeful redistribution. From year 2010 to 2100 the combined population of Europe and of Central and North America will have dropped from about 18% to 12% of the world's total, while the combined population of all other regions, which previously were less developed, will have risen from about 82% to 88%.

Figure 11 shows regional per-capita incomes and agricultural statistics in 2007. Nation-states are sorted into three economic classes and their differences compared. It is interesting to note that agricultural **GDPs** per total capita, were roughly equal in the USA, China and India, (415, 646 and 436 international dollars), but the amounts of agricultural land per agricultural worker were 182, 0.44 and 0.49 hectares respectively. The West continues to have fewer farm workers because it is more capable in other activities.

43

PART TWO

At present the entire world favors industrialism. Could it be that they know its potential to increase wealth is nearing an end? To modernize their own economies, today's eastern nations are following the same path as the west, again using industry as a first stage of diversification. In terms of world pollution this is the worst time in history. One of humanity's aims must be to slow this trend without depriving any nation of its fair share of development.

This speaks in favor of a more balanced world economy. There is no valid reason why industrial and agrarian cultures should not develop equally and share the same potential for expansion. In fact, agriculture is the more vital activity. Stopping industrial pollution is our first priority because of the need to slow global warming, but it does raise the issue of providing alternative work which doesn't pollute. I would go further and say that whenever possible we should provide jobs in agriculture which actually compensate for the pollution generated by others. If we arrive at a point where we have the luxury of choice in the distribution of effort, the priorities should be personal freedom and healthier living conditions first. The need to improve the health cf all people seems to be the more persuasive case.

Appendix 2 shows the variables in population. The **UN** estimated that there will be 8.73 billion people by year 2050, indicating a rate of growth of about 1% per annum compounded from year 2000. Developing countries still have the higher rates of growth, the poorest increasing 4.0 or 5.0% per year, but that will slow as the rate of food production approaches the world's maximum capacity.

An estimate of food supply related to world population is shown in **Figure 12**. It is based on the time taken to convert all marginal land to full use and increase the per capita food supply to solve existing malnutrition problems. <u>This goal would result in a maximum population of 21 billion in 750 years, and is used as the basis for the rest of the study.</u>

Figure 13 shows a possible redistribution of populations within this scenario from now until year 2750. Each region's growth would be proportional to its potential for increased agricultural output. Migrations in search of food are already happening in Africa and Asia whereas in the West, indigenous population growth is approaching zero or even negative rates. Therefore the rate of migration from East to West, which is already substantial, is likely to increase.

The reasons for the decline of family size in the West have been studied. Some factors are obvious, others not:

(1) Ageing populations reduce average fertility, but the average age will stabilize eventually.

(2) Women tend to seek paid employment when young, and are less likely to raise large families when older.

(3) Social traditions are expensive; the costs of education, health care, and living accommodation can be cited.

(4) In the last fifty years or so, there has been a reduction of male fertility in industrial nations due to chemical pollutants. (The fertility of other forms of life may also be affected, as explained below).

2.7 Lifestyle Hazards

Millennia of behavioral adaptations have made us old creatures living in a new and more dangerous world. Having exhausted the means of securing our physical safety we accept life as a gamble, raising our chance of survival by constant research. Human mortality can be listed under several headings but I've chosen just four:

Organic malfunction and cancers: 45% of all deaths.
Infectious diseases: 44%.
Starvation and malnutrition: 7%.(*)
Injuries, accidental and intentional: 4%.
(*) Actually, about 30% of the world's population has no reliable source of food and water and since this is linked to shorter life-span, deaths from the above-listed causes could be shared more evenly, perhaps with malnutrition as the biggest. Illness, poverty and malnutrition seem likely to continue at the same or higher rate. In affluent countries the provision of health care is unable to keep up with the demand. In poor countries it is unable to keep up with the need.

Most of the world is in a transitory stage between old agriculture and new industry. The result is pollution which reaches every region and causes biological damage on a global scale. It exists in the form of chemical imbalance, less nutritious foods, radiation of all kinds, noise, heat, artificial light in natural settings, non-biodegradable waste, synthesized chemicals used as food additives, preservatives, pesticides, herbicides and polymer modifiers some of which disrupt the bodily production of hormones. Every living thing is exposed to the pollution created by others and our species has exceeded them all in its output. Mammals and reptiles, fish and amphibians, were all biologically altered by swimming in water polluted by civilization before we even recognized the dangers to them and to ourselves.

Species with very short reproductive cycles may adapt to non-lethal amounts of pollution. by generational mutations. The greater harm is to animals higher in the food chain, with long reproductive cycles and more lifetime exposure to it. One concern is male fertility reduction which threatens all forms of life. Research on the effects of chemical pollution is proceeding slowly: About 30,000 chemicals have to be registered plus new ones as they occur. The **EU** regulation **EC 1907/2006** is called **REACH** which stands for the Registration, Evaluation, and Authorization of Chemicals. The European Chemicals Agency **(ECHA)**, maintains a database containing information on the properties of chemical substances. .

2.8 Conclusion of Part Two

The fact that the interdependence of species has continued for about a billion years seems to have little importance in human purpose. We prefer to use science to increase our comfort levels with a feeling of isolated security and with no thought for the long term consequences. As long as we use science this way we choose to live with danger, not necessarily from armed conflict but certainly from the toxic effects of scientific inventions on the environment.

In this era humanity will encounter its most difficult challenges. Achieving peace, comfort and security will be an uphill task requiring nothing less than allowing Earth to recover from its depleted state. Consider the much simpler need to reduce air pollution: The world took forty years to decide what should be done about it and still failed to achieve global participation. Governments were reluctant to cause economic disturbance and had little faith in the idea.

As recently as 1800, excessive pollution could have been avoided by continuing to use renewable clean energy from solar powered evaporators, windmills, sailing ships and water wheels. If it had taken a hundred years more to develop those old technologies the time would have been well-spent, and global warming would be minimal. But our ancestors were not well-informed and had no more chance to influence their futures than we as individuals do today. Whether they would have chosen otherwise with better foresight and organization is moot. However we should hope to be more informed in this day and age. There are three immediate tasks:

(1) Stop pollution without slowing the cash flow needed to pay for the transition.
(2) Determine how much more food can be grown, so that future population limits can be estimated.
(3) Develop a better and more equitable world through international cooperation.

In regard to equality the reason for economic growth is not to increase **GDP,** as such, but to grow cash-flow parallel to population. Sustainable effort may not be sufficient. Economies will always go through cycles of fast and slow growth. The best time to redistribute cash-flow is during fast growth and the easiest time to implement new initiatives is during slow growth.

The major change needed to see us safely through will be increased faith in our ability to reduce pollution. The world continues to hope for a better life and we should not underestimate the abilities of people to join in this common purpose.

In any democracy the wishes of the populace are most valuable when they are well informed. This is not a good time to be teaching the next generation that there are no limits to what humans can accomplish. It would be better for educators to stay within the bounds of known reality.

NATURAL ADAPTATION

Figure 7

THREE RANGES OF GOVERNMENT SPENDING SHOWN AS ANNUAL PER-CAPITA AMOUNTS OF INTERNATIONAL CURRENCY						
BUDGET ITEMS	**EXTENT OF ECONOMIC DEVELOPMENT**					
	COMPLETE		INTERMEDIATE		BEGINNING	
	As Currency	As % of GNP	As Currency	As % of GNP	As Currency	As % of GNP
Gross National Product (GNP)	30000	100	7000	100	2000	100
Cost of Government. Defense. Debts. Construction.	13500	45	4710	67.3	1600	80
Social Services, excluding Health Care.	12000	40	1750	25	300	15
National Health Care.	3000	10	440	6.3	100	5
Net Cost of other subsidized programs	1500	5	100	1.4	0	0
National ability to carry external debt in the course of business.	GOOD.		POOR		ZERO	

GNP Growth in 2005	2.8%	4.2%	6.3%
Population Growth in 2005	0.42%	0.68%	2.3%
Note: GNP is, GDP plus the value of exports minus the value of imports.			

Government Funding at Three Economic Levels

49

PART TWO

Figure 8

Years CATEGORY AND EXAMPLES	2005 Per Cap GDP [PPP] Dollars	05-06 Pop'ltn million	04-05 Exports billion $	02-05 External debt. billion $	02-05 Debt to Export values. Ratio	05-06 Growth of GDP % per Year	05-06 Growth of Pop'ltn % per Year	2005 Total GDP [PPP] billion $
WORLD	9362	6524.0	N.A.	N.A.	N.A.	4.70	1.14	61078
DEVELOPED	27172	1192.0	N.A.	N.A.	N.A.	2.80	0.42	32390
USA.	41800	298.4	9275.0	8837.0	9.528	3.50	0.19	12360
Canada.	34000	33.1	364.0	440.0	1.209	2.90	0.88	1114
Japan.	31500	27.4	550.0	1545.0	2.809	2.70	0.02	4108
European Union.	28100	457.0	1318.0	N.A.	N.A.	1.70	0.15	12180
TRANSITIONAL	9834	1013.0	N.A.	N.A.	N.A.	4.20	0.68	9962
Singapore.	28100	4.5	205.0	23.8	0.116	6.40	1.42	124
S. Korea.	20400	48.8	288.0	154.0	0.535	3.90	0.42	965
Russia.	11100	142.9	245.0	215.0	0.876	6.40	[-0.37]	1589
Mexico	10000	107.4	214.0	137.0	0.640	3.00	1.16	1067
DEVELOPING	4366	4319.0	N.A.	N.A.	N.A.	8.30	1.31	18726
China.	6800	1314.0	752.0	253.0	0.336	9.90	0.59	8859
India.	3300	1095.0	76.2	125.5	1.647	7.60	1.38	3611
Africa.	Af+Oth	888.0	Af+Oth	N.A.	N.A.	Af+Ot	2.30	Af+Ot
Others.	3275	1022.0	540.0	N.A.	N.A.	6.30	1.31	6256
HEAVILY INDEBTED	N.A.	N.A.	N.A.	N.A.	N.A.	N.A.	N.A.	N.A.
Central African Republic	1100	4.3	0.131	1.06	8.092	1.53	2.20	4.754
Sierra Leone.	800	6.0	0.185	1.61	8.703	6.30	2.30	4.921
Burundi.	700	8.09	0.052	1.20	23.077	1.10	3.70	5.654
Malawi.	600	13.01	0.364	3.29	9.030	[-3.0]	2.38	7.524

National Economic Comparisons 2002 to 2006

PART TWO

Figure 9

REGIONAL PER CAPITA GDP(PPP) AS PERCENTAGES OF WORLD AVERAGE								
Estimates For:	YEAR							
	1700	1820	1870	1913	1950	1973	1998	2005
Europe. (Approx. the Y2005 nations)	184	145	191	190	169	200	219	169
North and Central America	70	125	296	385	406	344	373	359
Asia (including Japan and the Middle East)	96	97	64	42	32	54	60	67
South America	14	18	24	48	83	70	80	99
Africa	48	45	42	33	40	30	23	23
Oceania (including N.Z. & Australia)	—	—	127	377	480	323	263	325

Per-Capita GDP(PPP) Distribution from 1700 to 2005

PART TWO

NATURAL ADAPTATION

Figure 10

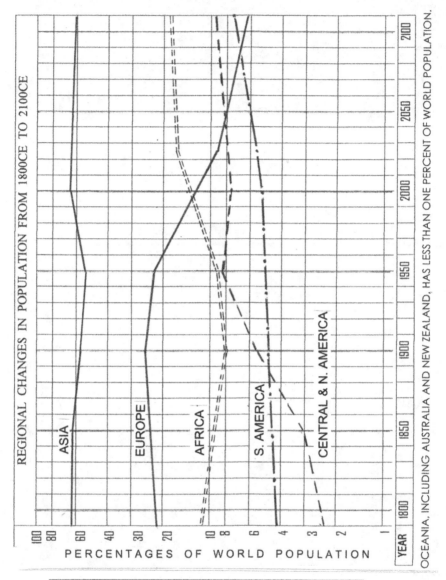

Regional populations from 1800 to 2100

PART TWO

NATURAL ADAPTATION

Figure 11

Columns	POPULATION millions	GDP per capita	FORESTS & WOODLANDS ha per capita	AGRICULTURAL LAND ha per capita	AGRICULTURAL GDP per capita	AGRICULTURAL GDP per agricultural worker	AGRICULTURAL LAND ha per agricultural worker	NON-AGRICULTURAL GDP per capita	NON-AGRICULTURAL GDP per non-agricultural worker
	1	2	3	4	5	6	7	8	9
World	6602	10000	0.63	0.820	3984	21917	4,51	6016	21944
Developed regions with a GINI index of 23 to 45									
EU	490	32900	0.50	1.600	591	29682	80.00	32309	67028
USA	301	46000	0.80	0.555	415	136165	182.00	45585	90532
Transitional regions with a GINI index of 40 to 55									
Russia	141	14600	5.67	0.877	679	11852	15.30	13921	29611
Developing regions with a GINI index of 30 to 75									
China	1321	5300	0.13	0.114	646	2475	0.44	4654	13580
India	1130	2700	0.06	0.135	436	1590	0.49	2264	12000
Africa	934	2800	0.96	1.050	415	1838	4.65	2385	16575
N. Africa	164	6072	0.00	0.360	436	7243	3,00	5636	26220
Sub-Sahara	770	2136	0.96	5.460	360	1415	5.00	1776	15490

GDP is in International Dollars, adjusted for Purchasing Power Parity [PPP].
GDP per capita is an indicator of national wealth, but only an approximate indicator of household incomes. Much of the GDP is used to support national infrastructures.
The GINI index is a measure of distribution; in these examples, related to personal income: Zero would be total equality. An index below 45 is considered 'good' but not necessarily an indication of wealth.

Incomes and Agricultural Assets in 2007

PART TWO

NATURAL ADAPTATION

Figure 12

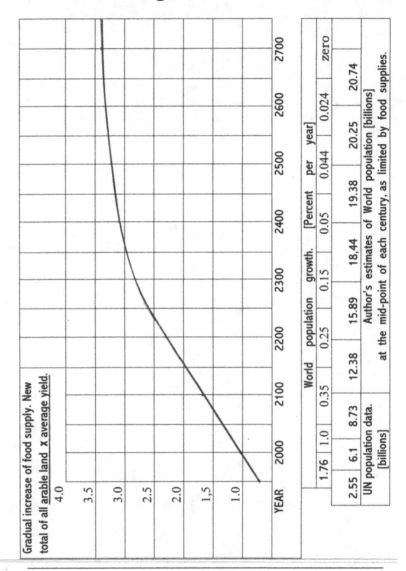

Food Supply and Population Growth

PART TWO

NATURAL ADAPTATION

Figure 13

YEAR	Africa	America. North & Central	America. South	Asia & Middle East	Australia N.Z. & Oceania	Europe	World total
2000	800	430	250	3632	28	760	6100
2005	858	450	270	4086	30	800	6524
2050	1400	660	400	5300	70	900	8730
2150	2300	1240	600	7100	220	920	12380
2250	3040	1900	800	8800	415	935	15890
2350	3800	2390	1200	9000	600	950	18440
2450	4200	3200	1600	8650	790	940	19380
2550	4600	3500	2200	8090	920	940	20250
2650	4700	3750	2500	7900	950	940	20740
2750	4700	3750	2500	7900	950	940	20740

Estimates of increased regional populations (millions), in proportion to increased arable lands and locally-grown foods, on the condition of no adverse effects from climate change.

Populations by Arable Land Increase 2000 to 2750

PART TWO

PART THREE

RATES OF CHANGE

PART THREE

PART THREE

3.1 Perspective

Reality is sensed through change and our concern is to find natural order in it. However, while the laws of Nature are immutable their results seem variable because we underestimate their complexity. For example, familiar events may be repeated for ever but many are beyond our control; they might seem familiar but in detail each one is unique. Our least amount of knowledge is in singular events and very long-term cycles of change. Such things have been studied, not because they might prove to be useful in our daily lives, but because historians have an innate curiosity and encourage us to follow. Our most-useful experiences come from modern history or the recent past, but for centuries, bits of natural history have been dug-up to assist further research. We use them to gain a sense of security from knowing how evolution might continue.

Though new ideas are challenging, we are accustomed to hearing about many paradigms that render 'old' theories obsolete. A sense of detachment grows as a type of self defense; that is, until we find ourselves involved in the consequences, which happens to be the case now, with the acceleration of global warming. In the months and years to come, we'll be more aware of our involvement. To understand the causes and predict their effects, we have to review the knowledge available in several disciplines, starting with the basics.

3.2 The Cost of Human Activity

260 years ago, when the world's population was only one ninth of today's, most nations believed in isolationism. Within 100 years the picture had changed. A wave of economic expansion started in Europe and international trade increased; it had a lot to do with market investment. Technology enabled people to travel faster and communicate in real time and the children of that era tried to turn science fiction into a new reality. In a few instances they succeeded and now, we awaken each morning to a more artificial world.

Since about 1970, the global community has been fully aware of the dangers of uncontrolled economic growth, yet it continues as before. Governments are not elected to attend to such things, only to provide jobs and maintain order. And so the world continues to follow its unlimited dreams. Human activities can be counted by the thousands but three categories are enough to give us a quick view of where we place our values. They are: Commerce and Services 62%, Industry 32% and Agriculture 6%, of **world GDP**.

Clearly agriculture is an undervalued activity but this is only the reflection of a bigger problem, which is our excessive interest in the other sectors and the way their growth is achieved. The service sector is largest in rich democracies where it can be as much as 90% of **national GDP**, while in poor countries it is as low as 20%. In contrast, the agricultural sector can be as low as 1% in rich countries while in poor ones it is as much as 50%. This imbalance is mostly due to the historic distribution of populations and resources. That is not to say new countries couldn't grow enough food, but simply that importing it enabled them to direct their efforts to more-profitable technological advantages.

RATES OF CHANGE

The trade value of everything finds its way into the **world's GDP,** which at the end of 2010 will be about 70 trillions of US dollars, predicted to grow at a rate of 6% per year for the next 5 years. One negative factor in GDP growth is the accumulation of national debts, which are often blamed on excessive household spending though they persists in all sectors of the economies of all nations. But there is a more serious factor which doesn't appear on any budget and will be unpaid as long as human activity remains unchecked. It is the '**natural debt**' owed to Earth's habitat. The entire populace carries both debts, though austerity is always transferred to the shoulders of the poor. Here are just a few of the early symptoms:

Pollution increases faster than population.
The biome changes and habitat is lost.
Conservation is talked about but not practiced.
Resource wastage increases.
Technical education is thought to be the solution.

If we leave financial debts unpaid for too long, lines of credit close one by one. Likewise, in natural debt, the resources on which we depend might disappear, causing humanity to be '**naturally poor**'. The parallel to our own world of economics is obvious, with one very important exception: Declaring natural bankruptcy will not help us to start anew.

No one knows what life will be like five hundred years from now, but we do know that Earth cannot bear a greater burden of misdirected effort without losing more of its habitats and species. We've exhausted the best possibilities, and if they cannot be restored our lives will be negatively affected.

The irony of technological development is that, far from causing the **HDI** of nations to converge, it is producing

67

climate change which threatens to make the most-needy populations its first casualties. Later in the book we'll see why. Eventually human conditions in the heavily populated tropics are likely to deteriorate.

3.3 The Modern Environment

As mentioned previously, the environment changes interactively with life. Exterminations only occur when changes are so rapid and extensive that they prevent cultures from adapting behaviorally, and biologically. The modern environment presents a challenge because our cultures recently evolved within a narrow range of conditions, and their further adaptation may only succeed within a limited amount of change.

Life still depends on the recycling of carbon dioxide while adding other elements as needed, but there is a destabilizing effect caused by our disproportionate emissions of carbon dioxide and other albedo-reducing green-house-effect gases (**GHGs**). This is exacerbated by our destruction of plant-life, which otherwise would grow to absorb some of the carbon dioxide excess. The result is that the average concentration of carbon dioxide in the air at ground level is increasing:

From 1000 to 1850CE it was 0.026 to 0.028%.
In 1950 it was 0.031%.
In 2010 it was 0.039%, and could double in 55 years.

In the 1980's, **GISS** and **UKMO** each reported on the relationship between carbon dioxide concentration and global warming, estimating that a concentration-increase from 0.03% to 0.09% would cause the world's average temperature to rise about 4 degrees Centigrade. Thus far, the unusually fast increase is causing climate redistribution in which some regions receive less precipitation while others receive significantly more.

RATES OF CHANGE

A sustained increase of only 2 degrees Centigrade would cause significant climatic problems. We have to look into distant history to find such a condition, but assessing the danger by using bits of data from tens of thousands of years ago is difficult because the only remaining evidence is from conditions which would be unthinkable today. Moreover, due to the way occurrences are measured (from the abundance and characteristics of life or from ice-core samples when available), there are areas of uncertainty in our knowledge. This is not to say that today's use of global temperature change as a proxy for environmental hazard is unreliable; it only points to uncertainty about timing and duration. We know there are chains of events with varying thresholds of sensitivity. For example, the melting of land ice requires only a small sustained temperature increase but it causes sea level to rise. This has altered ocean currents in the past, causing the redistribution of regional climates and eventually changing the condition of the entire world.

In regard to agriculture, large drought-areas may receive little or no benefit while monsoon rains could flood productive river deltas. Heavily populated tropical regions will be the most-affected. Elsewhere, the grain and soy yields from the main continental producers may drop 20% or more, not including further losses due to sea-level rise and flooding on coastal plains, deltas and wetlands. If the world has to feed a greater population while trying to eliminate malnutrition, these trends represent a global hazard. This will be reviewed later in connection with rainfall distribution.

Figure 14 shows temperature and sea level changes in the last 25 million years. At times, parts of the land now occupied by half of the world's population must have been covered with ice, in places over 1 kilometer thick.

PART THREE

The section of the timeline at the extreme right, marked 'Now', includes changes occurring in the last 20,000 years.

Our improved ability to examine relatively recent events, e.g. those occurring in the last 3 million years, makes it seem that sea levels and average temperatures are frequently reaching new extremes. Certainly polar temperatures were generally colder during the last 300,000 years, and the differences between high and low sea levels, and therefore water-ice transitions, increased within the last 200,000 years.

Ice ages are often thought to be periods of minimal change, but they are active. Sea-level changes may be delayed by a few hundred years as snow and ice repeatedly melt into the sea and then return as precipitation to cover the land. In other words, ice-ages deepen slowly but they can end abruptly. There could be another glacial cycle within the next 20,000 years or there might be none, but in the meantime Earth's air temperature is increasing.

The **biome** can be imagined as a living entity which reacts to changes in Earth's **albedo**: If it is lowered a healthy biome will produce more vegetation to raise it, and vice-versa. But when there is little or no vegetation to help in the recovery, any extreme state, hot or cold, could last a long time. The result is a damaged environment in which topsoil is washed away during thaws and eroded by wind when there are droughts.

In a recovery situation, oxygen and carbon dioxide tend toward specific concentrations related to the temporary conditions of living things, or conversely, those gases remain as life slowly adapts to their presence. Either way the World tends to approach a condition of active stability.

RATES OF CHANGE

Returning to the subject of the triggering effect of atmospheric carbon dioxide, **Figure 15** contains a list of other gases which also act as **GHGs**. It also shows their global warming potentials **(GWPs)** expressed as multiples of carbon dioxide. Mixtures of GHGs can be reported as **carbon dioxide equivalents** even though not all of them contain carbon. Atmospheric half lives are also shown. In the hypothetical case of a single GHG source which stops emitting while all else remains the same, the time taken for its atmospheric concentration to drop to one tenth of its original amount would be 3.33 times its half-life. Of course reality is more complex because all things are interactive.

The counterpart of GHG half-life is the hysteresis between a new emission and its related temperature increase: As the atmosphere's mass gathers solar energy and as life tries to adapt to each increment, the temperature takes time to equilibrate and due to normal fluctuations that moment is difficult to determine. The only reliable measurements are averages of change within periods of time. For example, warming was 0.7 degrees of Centigrade in the period from 1980 to 2009. Past temperature measurements show that it cannot have been rising at that rate for very long and that it is still increasing due to earlier accumulations of GHG.

We are still far from having controlled emissions. If or when we are able to stop them, the amounts already in the atmosphere will continue the warming process until they disperse, and the cooling period begins.

Figure 16 is a table showing large changes in world temperature within the last 20,000 years and their rates of change. The three peak rates were as follows:

71

PART THREE

(1) An increase of 2 degrees of Centigrade in 150 years at the time of the inundation of the Mediterranean and Black Seas about 10,500YA. World population estimate: 3 million.

(2) A decrease of 0.7 degrees in 50 years at the beginning of the Little Ice Age about 600YA. World population estimate: 600 million.

(3) An increase of 0.7 degrees of Centigrade in the last 25 years which could mark the beginning of the end of the industrialized era. World population increased from 4.8 to 6.8 billion.

These changes were relatively small. Fast change is not always due to atmospheric **GHG**, and whatever the cause it may not wait for life to adapt. The rate of change in the last 25 years was about twice that in item (2) but thus far has lasted only one sixth as long as the more extensive in item (1).

Even so the global rate of change is increasing and its duration will depend on our ability to eliminate the cause. For safety it would be preferable to stop unnatural global warming by reducing the carbon dioxide concentration to the pre-industrial level of 0.025%, but as we shall see, that won't be possible.

As to our vulnerable condition, I should point out that a prolonged warm preceded the **Little Ice-age** of the fourteenth century, causing an outbreak of bubonic plague which was exacerbated by food shortages. In recorded history this and similar events may have caused over 100 million human deaths. It shows that a combination of population crowding and changing climate greatly increases mortality. Today's population is about eleven times what it was in the fourteenth century and is now much closer to Earth's limited ability to support us.

RATES OF CHANGE

3.4 Temperature and Sea Level

Arctic and Antarctic ice sheets are the last remains of previous ice-ages. Indications are that a new extreme of ice-melt could break the **Neogene's** ice-age cycles by reducing Earth's **albedo** below a critical limit. **Figure 17** contains two graphs, the upper showing average surface air temperatures in the 25,000 years before 2005 and the lower showing sea levels. The **IPCC Fourth Assessment Report (2007)** described several possible ranges of future temperature.

At civilization's present stage of maximum growth, the ice-melt will be one of the destructive elements of global warming but it will be gradual and the timeline of serious flooding is uncertain. Moreover there is the prospect of our subsequent adaptation to warmer climates. The first stage, unfortunately already in progress, is the destruction of the habitat of polar species and therefore reduced options for human settlements along the polar coasts. In the longer term, rising sea level will affect all maritime populations. We depend on agricultural tracts close to sea-level and though annual flooding is already considered normal in many regions, the main concerns are the amount of land and other capital assets that will be permanently inundated. Not every country will find it possible to protect itself with sea barriers.

The recent losses of life in tsunamis and storm-surges clearly shows that people and vital infrastructure are not safely above sea-level. By my estimate, about 40 million people live or work less than a meter above high tide and about 400 million are within 10 meters of it. Today, Earth's land-ice volume is about 32 million cubic kilometers, an amount which could produce 65 meters of sea-level rise, or 2.61 meters for each million

73

cubic kilometers of melt. The fastest melting ice, in Greenland and the West Antarctic, might add up to 9 meters of sea level while the slower melting East Antarctic could produce 54 meters. The remainder would come from the continued dehydration of continental highlands and the loss of forests.

However, global thaws are rare occurrences and it is reasonable to expect not more than 15% of all land-ice will melt in the next 2000 years. Hopefully the rate of melting will peak in 500 years then slow to zero at a sea level no more than 10 meters higher than today's (again, my estimate). During the sea-level rise, environmentally-important wetlands could turn into unusable tidal swamps while sea-ports, container terminals and power plants would have to be relocated, and sea bridges built to ease transportation and save fuel.

3.5 The Total Effect

In less than 20.000 years (0.0005% of Earth's total age), the **anthropogenic effect** has grown from that of a rare but inventive species of primate to an organized force which has literally changed the face of the planet. Here are some statistics showing the great increase of human forcing on the natural world.

(1) About 83% of anthropogenic carbon-emission is due to the use of fossil fuels. Only 17% is due to the metabolic use of energy derived from foods.

(2) Since humankind changed its activities from hunting and gathering to mining and manufacturing, world population has increased one thousand fold.

(3) The world's *per capita rate* of energy usage increased thirty fold.

(4) On the basis of (2) and (3), humankind's energy consumption is 30,000 times larger than it was 5,000 years ago.

(5) The anthropogenic effect has grown even faster than stated above. Our use of metabolic energy had a usable output of about 18% of intake. Today we use processes which have thermal efficiencies of about 50%, and adding this factor causes human activity to have at least 80,000 times its previous effect. Only 10% of the energy is used for agriculture, 90% goes to industrial and social growth.

There are too many regional effects to provide a complete list. Here are just a few:

(6) Population and activity, increase urban development, environmental pollution, wasted materials and the use of synthesized chemicals. Deforestation, soil erosion and the destruction of species continues faster than our population growth.

(7) Within decades there has been an increase in the frequency and energy level of storm systems in the sub-tropics and temperate zones, causing more property damage from storm surge, flash floods and mud slides.

(8) Regions with coastal mountain ranges are losing vegetation as rain clouds move across at higher altitudes.

(9) Spring flooding is aggravated by more river ice-floe blockages caused by faster thaws. Water catchments previously silted, now overflow. There is a reduction of water supplied to reservoirs later in the year and warmer summers increase the need for farm irrigation.

(10) Disease-carrying insects and plants from the tropics are harming plants and animals in other zones.

(11) Forests and woodlands contain more dead trees and dry undergrowth, which raises the fire hazard.

Note: **UNEP** *issues reports, on the worldwide efforts of interested parties to conserve biological diversity.*

We have attempted to mitigate the above-listed effects without reducing the causes, but the global result is unpredictable and likely to do no more than replace one set of problems with another. Some attempts are admittedly controversial e.g. the genetic modification of plants and foodstuffs to increase farm productivity and cloud seeding to cause precipitation where there is insufficient fresh water.

There are other ideas such as fertilizing the oceans to grow more phytoplankton in an effort to accelerate the removal carbon dioxide from the atmosphere. But new information on the state of the oceans suggests that marine life is already going through its own metamorphosis in response to higher acidity. In November **2008,** the **UN IMO** issued a moratorium on ocean fertilization pending further investigation but experiments at sea are continuing. Even more-radical ideas have been offered for reducing global warming but it seems they will remain in the realm of science fiction.

I regret to say that such efforts seem typical of uncontrolled development. Trying to find shortcuts which fail to address the basic issues only encourages us to continue polluting.

3.6 Energy Resources

Figure 18 shows the main hydrocarbon fuels in use today, their potential heat energies in gigajoules per tonne (GJ/t), and their carbon content in tonnes per gigajoule (t/GJ).

There is little agreement on the time when the first use of combustion energy occurred, but for warmth and cooking it might have started about 100,000YA using wood for fuel. Much later, charcoal and plant oils supplemented wood and this practice grew with the population until demand exceeded supply and fuel gathering became a daily task; in the later stages, charcoal production was a significant factor in the depletion of forests. There is still a market for charcoal in rural areas but it is becoming scarce. The demand for metal implements increased the use of coal and this gradually replaced the bulk of charcoal usage. Coal is still the most-used of all fossil fuels. The figure shows that it is about 60% cleaner-burning than wood but 60% dirtier than oil.

As previously mentioned, apart from causing vast amounts of environmental damage, there are two categories of carbon dioxide emissions that we have greatly increased:

The first is the biological or metabolic component. The human body emits about 0.23 tonnes of **GHG** as carbon dioxide ($CO2$) per year. Our 2010 population, of 6.8 billions, emits a total of about 1.6Gt of $CO2$ per year, yet if population remained constant and the only activities were to grow and eat food, our species could be called carbon-neutral.

However, in addition to increased population we are involved with converting all of Earth's resources to support further economic growth. Therefore the second component is purposeful GHG emissions. This category includes the use of fossil fuel and non-renewable biomass to obtain energy for all the developments of civilization. In addition there are emissions from agricultural fires, cement and petrochemical production, and the decomposition of wasted resources.

The net total of emissions from all sources, after the benefit of biological sorption, was about 23Gt of carbon dioxide equivalent in 2009 and following the present trend, it will be 40Gt per year in 50 years or less.

Species find it easier to adapt when others on which they depend are acclimatized, yet we continue to destroy habitats. The best environment for life is one with an abundance of healthy foliage yet it seems we intend to find a way to survive without this benefit. The **anthropogenic effect** includes the willful destruction of forests and agricultural land, which are the very things we need to help clean the atmosphere and reduce global warming. If we were to allow enough time we would see the plant population increase to take more carbon from the atmosphere but our rate of GHG emissions is approaching critical proportions; climate redistribution will decimate some plant species before they can fully adapt. The destruction of northern pines by the pine beetle due to a moderate climate change, is an example of what could happen in future.

The Carbon Cycle
Figure 19 shows the carbon-flow lines. In equilibrium there would be only the seasonal ebb and flow in the **biome** due to the fact that most of Earth's land mass is in one hemisphere. The land and its life-forms would then recycle just over half of the atmospheric carbon and the sea and its life-forms would account for the remainder. Carbon buried in rock formations or transported to the sea-floor can stay out of the cycle for eons unless it is released as oil, coal or gas, in which case it is brought to the surface and used for fuel or transformed into other petrochemicals.

Clean Energy
Because anthropogenic warming is already in progress, we may not be able to stop climate change in the next 50 years, but by reducing air pollution and habitat

destruction we could aim to moderate its global effect and prevent some of the harm.

Perhaps the most important consideration is how much longer fossil fuel reserves will last. In its **Statistical Review of World Energy (June 2010), BP** presented usage estimates and reserves which seemed to indicate about 120 years for coal, 60 for natural gas and 45 for oil (based on year 2009 usage rates). However those life-estimates might increase due to continued exploration; history shows that known reserves increase approximately in proportion to our increased usage. But they may finally start to fall due to the cumulative effects of population growth, economic expansion, and a lack of sufficient clean energy to replace them. Thinking in terms of the distant future, it seems unwise to plan on using every last bit of reserves. Who knows what humanity will need for heating or for hydrocarbons to feed petrochemical development?

Coal will remain the biggest polluter: More than 1000 coal-fired power plants are under construction or planned. No doubt they will be more efficient and emit slightly less pollution per KWH, but electricity gathered from clean sources would avoid the pollution problem altogether. High efficiency in hydro electric generators and wind farms is only needed to reduce costs; there are no combustion energy losses and even when working at low efficiency they have no **GHG** emissions.

A particular advantage of renewable clean energy is that small generators can often be installed locally, reducing long distance transmission losses. This infrastructure distribution also offers more security than having fewer and larger power-generating complexes. It is already possible for operators to feed surplus power into regional grids. To quantify the future emissions-reduction potential we should look at the different ways energy is used and consider the local prospects in each application, but that would be a very detailed study. Our present capability is limited to global statistics from national and international energy providers.

These sources give ranges of data which are in reasonable agreement. In 2009, world fossil fuel consumption was equal to about 407EJ of primary energy. Another 57EJ came from the mature carbon-free technologies of nuclear, hydro and geothermal power, and a further 8EJ came from bio-fuel, wind, solar and tidal power (bio-fuel is considered to be carbon neutral, the others are carbon free).

With this information one can calculate that emissions would have been about 16% higher without carbon-free energy. The calculation involved three adjustments and a mixture of pluses and minuses: Firstly, the often advertised design capacities of clean energy systems cannot be used directly since on average they run at about 25% capacity. Secondly, about 98% of their output is in the form of electricity, which is ideal for clean energy generators, whereas using fossil fuel we would have burned twice as much as the equivalent energy delivered, due to the inefficiency of combustion–to-electricity convertors. In other words, clean energy saves double the amount of emissions from the fossil fuels used to produce electricity. Thirdly, about 40% of our energy usage is non-electric, mostly in transportation, civic, industrial and residential applications.

The long term solution will be to convert everything to use clean electricity except for the few heating applications where solar radiation and geothermal energy can be used directly. This gives us another reason to develop all types of renewable energy at the same time.

For many reasons we have to raise the status of the **clean energy initiative** from an opportunity to an emergency. It takes time, and there is every indication that global warming will accelerate and become a severe danger if we cannot stop fuelling it in the next 40 or 50 years. To mitigate global warming we have to reduce coal usage and increase clean energy as quickly as possible, reaching total replacement long before the world's population reaches its maximum number.

Based on recent history there will be no need to increase *per capita* energy use; the total energy available need only be increased in proportion to population. But we should aim to supply all of it (1500EJ of energy per year) from clean sources. A quadrupling of the present total of designed capacity would be sufficient. Progress should be such that we achieve 90% of this by year 2050 and 100% by 2300. It will be even more difficult than it seems:

Only 14% of our usage is clean energy and 70% of that comes from the mature technologies of nuclear, hydro and geothermal power. If the future increase depends on new technology alone it will have to increase its output about 75 fold, and if we exclude bio-fuels from long term plans because the land it uses now will be needed to grow food, the other new technologies will have to increase 115 fold. Clearly we have a long way to go before clean energy can meet all future demands. Of the small amount now available, 90% comes from 'wind farms' but for reasons already stated we need to develop a mixture of clean energy sources.

Therefore we have to accelerate the use of viable systems in a way that produces the desired result in the shortest possible time. The budget should be at least 1% of **world GDP** in each of the next 50 years.

Renewable Bio-Fuels

As commonly used, the term 'bio-fuel' implies 'clean energy', but they are two different things. Bio-fuel production should be seen as a temporary way of increasing clean energy. Its claim to be carbon-neutral is on the basis that emissions from its combustion are removed from the atmosphere by fuel crops still in the growth stage. However, if bio-fuel crops *replace* other plants in the biome which absorb carbon without being burned, the claim seems to be unproven. The atmosphere already contains enough carbon to support plant life. The full benefit of bio-fuel is only obtained if it *increases* the biome's plant life. This means it would have to be grown on land which otherwise would have contained no plants. It seems the land now used to grow bio-fuels should be returned to agriculture if possible. This assumes that the ancillary processes of agriculture and fuel-cropping are equal in regard to their own emissions.

Farmers today can obtain high profits from bio-fuel crops, especially those used to supplement gasoline, and this is raising food prices. Alcohol fuels can be produced from crops such as beets, corn and sugar cane, while bio-diesel comes from crops which yield oils. Together they represent less than 1% of the world's non-fossil energy, yet they've had a negative effect on world food supplies. In 2008 the price of cereal grain rose sharply and is still rising, causing severe problems for low-income populations. An **IFPRI** report, **'The World Food Situation' (December 2007)**, explained this bio-fuel effect in some detail.

A report by the **USDA**, **'Estimating the Net Energy Balance of Corn Ethanol' (1995)**, dealt with the topic of delivered or net energy, versus gross-energy. It found that the amount expended in growing corn and obtaining ethanol from it was about four times the amount delivered to the point of use. Other countries have improved the economics using the woody waste from sugar cane instead of burning it.

Aviation might be difficult to convert to bio-fuel because practically all of its usage is in kerosene-type fractions of oil. Eventually a jet-fuel shortage could be eased by producing bio-fuel from soy, canola, or sunflower seed. The cost per unit of energy, including subsidies and other donations, would be high by today's standards but even so, aero-engines using bio-fuel have been flight tested. Therefore the options are either to continue using fossil kerosene on a controlled basis, or to produce a light bio-diesel from crops grown on agricultural land. It seems the amount could be limited to about 2 or 3% of the worlds total energy usage, though food supplies would be seriously affected.

3.7 Rate of Energy Change

Energy-change estimates are about efforts to reduce pollution while increasing total energy usage. There is no comprehensive world study at this time. Published reports deal with commercial interests but globally the chances of rapid development can only be estimated. The unbiased comments in italics are my own:

- In its report **'International Energy Outlook 2010'**, the **US DOE** estimated world energy usage would increase 49% from year 2007 to 2035.

  Apparently this refers to marketed energy, without hydro or recent developments in renewable energy. It seems to be increasing about twice as fast as world population, which would raise carbon dioxide concentration to a critical amount by 2050. Since there is, as yet, no world intention to restrict total energy usage, global warming could exceed a safe limit.

- In '**World Energy Outlook 2009**', the **IEA** reported that primary energy demand was projected to be 16800 Million tonnes of oil equivalent in 2030.

  Reference tables show that one tonne of oil equivalent could supply 42.53 million BTU. The primary energy demand in 2030 would then be 714 QBTU.

- In '**Global Wind Energy Outlook 2010**', **GWEC** reported that they hoped to exceed an IEA projection of 573 GW for installed capacity by the end of 2030.

  Our calculation shows that, with a typical capacity factor of 25%, 573 GW as installed would supply 1255 TWh of electricity per year. If that amount of electricity were produced from coal fired power stations it would consume 213QBTU of primary energy per year. Depending on one's interpretation of the data, this shows it is hoped that in 2030 wind energy will reduce GHG emission by about 30% from what it might otherwise have been.

- In '**Concentrating Solar Power. Outlook 2009**', **Greenpeace International, Solar PACES** and **ESTELA**, reported 436 MW capacity from solar thermal collectors installed by the end of 2008, a potential to supply 7% of the world's power needs by 2030, and fully one quarter by 2050.

- In '**Trends in Photovoltaic Applications**', the **IEA** reported that 6.2 GW of PV capacity was installed in 2009 bringing the total thus far to 20.4 GW.

Note: As mentioned previously the electrical output of clean energy generators is usually given as 25% of their designed capacity. This is due to operating conditions which may change annually and daily.

Great technical advances have been made to enable us to replace fossil fuels with renewable clean energy, and the rate of added capacity is accelerating. But this excellent progress may be too late to replace all, or nearly all combustion fuels before global warming becomes a serious problem. There should be a limit on total energy usage so that clean energy may be truly used to reduce emissions, and not just used to expand the economy even faster.

3.8 The Crisis

The general outlook is that the extra activity required to solve the already existing problems of deprivation and malnutrition, will accelerate emissions and climate change because we have very little clean energy to work with at this time. Either way, world population will increase until the consequences of further growth become too severe. It is civilization's responsibility to ensure that the transition is gradual and asymptotic to a final controlled condition.

Figure 20 contains three graphs based on probable energy usage and showing the net amounts of carbon dioxide emissions per year, the state of the **biome,** and the carbon-dioxide concentrations from year 1850 until now, projected to 2650. It assumes emissions will finally approach a maximum related to maximum population in 2650 (ref. **Figure 12**), after the present excessive emissions have been quelled and controlled. But we have only 50 years in which to achieve 90% clean energy (in this example, started in year 2000). Subsequently it relies on cultural evolution toward a more agrarian world with restricted energy usage. Living conditions will continue to deteriorate until rapid global warming has passed, at or about year 2200. I'll describe this timeline step by step:

The main feature of the top graph is the high rate of emissions in the period from 1850 to 2250. Except for this, the graph follows a gradual rise which is consistent with population growth. In other words the emissions anomaly is purely the result of excessive growth and per-capita activity in the last 200 years. By year 2070 the total net emissions, from all sources will reach about 37Gt of CO_2 per year. Then the graph shows a hoped-for result that energy usage will be cleaner and emissions will subside. However climate change with its further loss of plant life, and the further anthropogenic destruction of green habitat will decrease the **biome**'s effective rate of carbon sorption and sequestration until year 2200.

When clean energy nears its highest percentage of total energy used, fuel emissions may fall to about 4Gt of CO_2 per year. Even so, the combined net emissions from all sources, stays above 18 Gt of CO_2 per year. Agrarian lifestyles and habitat initiatives might succeed in limiting the atmospheric concentration to 720ppmv.

Beyond year 2300, net emissions follow a rising trend caused by continued population growth, but atmospheric carbon dioxide is gradually reduced due to the increase of biological sorption, and more non-biological adsorption as the average temperature of air at ground level falls slightly.

By 2650 the total net emission rate will be 27Gt of CO_2 per year. Population growth stops only when all suitable land is used for agriculture, at which stage there will be the possibility of limiting atmospheric **GHG** to 600ppmv of carbon dioxide equivalent (allowing for normal fluctuations) while stabilizing the average temperature at 4 degrees of Centigrade above the 1950 level.

3.9 An Agrarian Scenario

The ultimate effect of all activities should be carbon neutral. There is no escaping the fact that our lives will soon depend more on agriculture and less on everything else. If it is well planned there will be less pollution and more accessible work for all.

With the collaboration of the **FAO**, the representatives of Heads of State attend 'Conferences on World Food Security'. A **'Declaration of the World Summit on Food Security'** was issued in **Nov. 2009** stating four strategic objectives. Briefly, these were: to fully reach the target of **MDG goal 1**, implement the reform of the **FAO CFS**, reverse the decline in agricultural funding, and face the challenges of climate change with particular concern for small producers and vulnerable populations,

These benefits have been difficult to achieve, as the economies of nations are so unequal. It seems to me that **cultural evolution**, which obviously includes agriculture, should aim to increase equality by reducing non-vital industry and increasing agricultural jobs. This would be a greener solution to global warming than, for example, awarding carbon credits for growing bio-fuels, though even that may stop voluntarily when world population is larger and food becomes the only concern.

Any plan starts as a type of subsidy for poor nations, but in the context of reducing climate change, it has to be aligned with our need to stop non-essential activity and increase agricultural employment. This should be combined with better health care, housing assistance and training in essential occupations. Such basic humanitarian provisions could help the majority of people who thus far, have been unable to obtain them. This will be described in more detail later in the book.

PART THREE

A change of this magnitude will take decades to accomplish. National governments will have to plan and implement their economic revision, probably with regional experts advising on the best ways to contribute to the global effort, and especially by focusing on agricultural essentials such as water, power, transportation and the cultivation of new farmlands.

The actions for success can be summarized as follows:

- It will be essential to increase food output faster than population, even if the lower yields per hectare of new farm land make it difficult to achieve.

- Humanitarian services, energy supplies, raw materials and credit facilities, for the larger population, will have to be provided.

- A temporary increase in emissions, due to new-construction activities, will be unavoidable, but clean energy must be developed faster to minimize the risk.

There will probably be an increase in the planning and construction of new towns, located wherever there is a potential for agriculture and clean energy. This would reduce unemployment while avoiding the further crowding of old urban centers. Readers will recognize that satellite-towns already exist as urban sprawl, but not yet for increasing agriculture by using clean energy from local sources.

The quadrant diagram, **Figure 21**, shows how four negative influences might combine to increase the natural debt and make it difficult to extract a living from available land. Their characteristics are as follows:

RATES OF CHANGE

(1) If we assume that agricultural land should increase in direct proportion to population, it will be grossly underestimated.

(2) Unpredictable activities grow with increasing population.

(3) Per-capita pollution increases with new construction.

(4) Because of climate change, first estimates of the amount of farmland needed will have to be increased.

The goal is to reduce adverse factors and try to make the first land estimate sufficient. If we see the world as a single state, this amount might be determined from the graph in **Figure 12** otherwise each region or trading-collective would need its own graph. Actually, the latter seems to be the case, since food is one of the resources with varied international trade; some countries have almost no agriculture of their own while others are involved in exchanges involving a variety of food items. Nevertheless the world-state example is true in principle, and global problems affect everyone eventually.

Year 2007 inventories of arable and potentially-arable land showed that 14.73 million square kilometers could ultimately be increased to about 40.8 million (see **Figure 22**), or about 2.77 times the amount. A previous **UN** estimate showed 14 million could be increased to 50.2 million, but that was revised due to the future effects of global warming and losses of coastal land associated with rising sea-level: It could be revised again. The operating mode suggested here, is that nation-states would not grow food solely for themselves, nor would they be required to grow a surplus, but those with excess capacity would have the option to export to the world's neediest regions and gain credits for it.

Using year 2001 as a starting point, world population is expected to rise by a factor of 3.44, and since adequate nutrition calls for 50% more food per capita, the world's rate of food supply at maximum population would have to be 5.16 times higher. Then, the world average crop yield per unit of area would have to be proportional to 5.16 divided by 2.77, showing an 86.3% increase on 2001. That will be difficult to attain because at the time of maximum population about two thirds of all agricultural land will have been reclaimed from neglected or marginal areas, and even traditional farming regions will be challenged by climate change.

Eventually, efforts to get higher yields from existing farms will have run their course. Genetically modified organisms (**GMOs**) are already in the food supply producing higher yields. According to one report, 22 million farmers planted genetically modified seed in year 2008, and the **UN FAO** is reported to have taken an advisory role to inform governments on safety and efficacy.

It seems to me that in view of the need to increase farmland, the excessive use of GMOs to provide food only passes the hard work of cultivating more land, onto future generations who will have less time remaining in which to accomplish it. A more humanitarian option would be to employ more workers per hectare now, and develop small-farm cooperatives supplied with water and other services. Obviously there will also be opportunities to employ more workers on infrastructure projects, and this should be recognized as another opportunity to reduce poverty and hunger. If economic growth were equitably shared between agriculture and industry, these could be considered permanent jobs with a good future.

Taking this idea further, as part of a scenario of adaptation there should already be a plan for new farmlands. There is no immediate shortage of unprepared land, but due to low or negative profit margins countries have been reluctant to develop it. As with the increased use of **GMOs**, this only postpones the problem, leaving it for future generations to deal with.

Fresh-Water Supplies

Finding enough fresh water has always been a priority; population centers are already located near essential supplies or are trying to survive as best they can.

About 5000 years ago, large cultures found they had the extra task of ensuring they had enough fresh water for contingencies, and in time this caused them to search for reservoirs and build aqueducts. Today, as populations grow faster and climate redistribution threatens, the search for water is again urgent, especially in regions which over the years have experienced drought. There .are many examples of this condition in the Middle East, North Africa, East Asia and areas which happen to be in the rain-shadow of mountain ranges. In our soon-to-be overpopulated world, where can populations move without causing more shortages? Only where water can be delivered or transported.

The **UN FAO** provides statistics in its database called **AQUASTAT**, and **UNEP** promotes research and gathers data from around the world. From readings of this I would say that the difficulty of improving the water supply in good time is second only to that of switching the world on to clean energy. The problem has its own dynamic: Fresh water supplies are generally in convenient river basins, lakes or shallow aquifers, with limited withdrawal capacities. Wetland agriculture around the world is threatened by climate change as rivers become dry in one season and wash away topsoil in another. In drought stricken regions, wells are drilled into fossil aquifers, as deep as 1000 meters, though they are not considered rechargeable. Even with these extreme measures, rural communities are disappearing. Without a plan to increase water distribution world agriculture may reach its upper limit in 50 years.

Figure 23 shows further examples of water depletion and its visible signs. Clearly this is not merely restricting growth but is destroying what we have built.

As usual, the task is easier to explain if we consider it from the global point of view: Logically, the world's population growth will slow and stop at some time, and we have to be certain there will be enough food and water to support that number wherever they live.

As noted previously, the availability of food will be limited to 5.16 times the amount produced now. At first sight, the easiest plan would be to convert pasture lands or forests to food cropping, but this depletes the mature plant population which is already stressed. A more sustainable balance could be achieved by restoring desolate land with water diverted from precipitation run-off in areas having a plentiful supply. Contrary to first impressions, this doesn't deplete the world's water but only diverts its flow before it returns to the sea and contributes to precipitation. We should not divert water from any area if that stops plant life. The real difficulty will depend on the extent of climate change and the distance of water transportation. Therefore our best efforts to limit global warming are required now while there is time and before measures become crisis responses.

Figure 24 shows the world's water distribution by latitudinal zones in year 2010, and in two possible trends, 'A' and 'B' at years 2060 and 2200 respectively. (**GHG** concentrations were obtained from **Figure 20**):

The world's annual precipitation on land is about 120,000 cubic kilometers. Normal evaporation plus transpiration, which I call 'total evaporation', removes about 75,000 cubic kilometers leaving 45,000 cubic kilometers of land-water-flow (**LWF**) to find its way to the sea. We already divert about 4380 cubic kilometers of water flow annually (**DWF**), to use before it returns to the sea. Studies of the world's hydrological cycle show a range of estimates so we can only use approximate averages to show Earth's capacity to support the main populations:

Trend 'A' is an optimistic view. Some rainfall will move from the tropics to semi tropical and temperate zones. Depending on our ability to use 40% clean energy before 2060, living conditions would be recoverable after global warming, though with great difficulty in the tropics as it will call for about 50% of tropical **LWF** to be diverted by 2200. Assuming that the previous plateau in year 2060 was achieved, the additional amount needed in 2200 would be 1343 cubic kilometers per year for the northern and southern tropics combined. I could not imagine any fresh water production using desalination or other process capable of supplying more than 10% of that short-fall. The remainder would have to be delivered through long distance pipelines.

Trend 'B' is the consequence of failing to use 40% clean energy before 2060. Neither the world's average temperature nor the increased need for water, are predictable after year 2060.

All of this relates to the future of our food supply which has many unknowns. They serve to emphasize the need for this investigation:

At this moment we cannot be sure of the future ability of temperate zone farmlands to feed the rest of the world, nor of any redistribution of world population in response to climate change. Stopping the present depletion of the world's green vegetation would be a hoped-for sign but total restoration of the biome seems to be far into an uncertain future.

An interesting observation is that, in theory at least, desert countries which totally rely on desalination need only increase their capacities in proportion to population to have sufficient water. This assumes that enough food can be imported without the need for domestic hydroponic agriculture, but that too would depend on the world food situation and global warming.

Other unknowns are the side effects of increasing the **DWF** from 10% of **LWF** to 26%, yet in 150 years we shall need it. There must be planned flexibility in our redistribution of water.

Consider the plight of Beijing, China. In that region 33 million people are waiting for the South-North Yangtze River Diversion to be completed. By 2050 it will deliver 45 cubic kilometers of fresh water per year for industry and farmlands, increasing the world's distribution capacity by only 1%. The project budget is estimated to be 65 billion US dollars and might go higher. Using this as a guide, the future global cost of increasing water distribution might average 3 to 5 billion dollars (in 2005 currency) for each cubic kilometer per year capacity.

94

RATES OF CHANGE

China's **DWF** will also gain 50 cubic kilometers per year from cloud-seeding technology. Some will return directly to the sea and some will help agriculture. The world may need one hundred times that increase, not as new rainfall but delivered through pipelines.

In total, China's fresh-water initiative will be the most expansive of all as it seems to include drawing more water from the Himalayas. India has a similar population but a smaller land area and different topography. It plans many smaller projects closer to the population centers and combining water supply and hydro-electricity. Sub-Saharan Africa could use more of its own fresh water for irrigation and in this regard is similar to water–rich countries in Eastern South America. Africa's existing water problems are variously attributed to its increasing population, poverty, and insufficient planning. There are many wetland areas which support local populations though supplies are not reliable everywhere.

As a general rule, more reservoirs will have to be constructed in all countries, and older systems will have to be repaired and enlarged. Recycling water to reduce distribution costs will be more important than ever, but even without it we can be sure that Earth has its own recycling system with the sea as its reservoir.

Understandably there are factions opposed to changing natural water courses, yet at this stage we cannot move cities or relocate billions of people, and in any event water has to be delivered wherever there is a potential for agriculture. The increasing international transfer of **'virtual water'** in the form of food is a clear sign of the magnitude and scope of this problem. (Refer to **UNESCO. Water. A Shared Responsibility. Second World Water Development Report. 2006).**

PART THREE

Farm Development

Today, farm development consists mainly of changing crops, improving yields and modernizing processes. The cost of actually buying, converting and cultivating land depends on local conditions and is unknown as a global average. Certainly this will require more infrastructure, agricultural equipment and farm workers with higher rates of pay per unit of yield. But from a global perspective, these should be balanced by planned reductions in non-essential activities in other sectors of the world economy.

The all-inclusive cost of providing safeguards against regional water shortages, especially with the approach of climate change, will be about 25% of one year's **world GDP**, spread over the construction period, followed by 0.3% of world GDP spent annually for upkeep and modifications.

3.10 Conclusion of Part Three

We are halfway through this study of cultural evolution and it seems the journey of mankind will complete a circuit, taking it back to an earlier age of relatively harmless agriculture. However the resemblance is superficial. We cannot go back and as yet we have no organized plan for the future. Switching to a mainly agrarian economy is suggested as a way to deal with the present crisis, the overriding goal being to establish conditions which will allow each generation to support itself without borrowing from the future. I cannot see any reason why most of the value now ascribed to industrial output, should not be transferred to food output. It could be a win-win situation: Even if agrarianism proved to be merely a temporary phase in cultural evolution, at least our descendants might be spared from the worst effects of climate change.

RATES OF CHANGE

Agro-industrialism using clean energy, will not mean the end of industry but it will be a less hazardous world, using more workers and reducing unnecessary pollution. Such purposeful employment need not deprive us of rest and recreation, in fact I imagine it will be a more satisfying and stress-free lifestyle, but there will be an era of austerity until new patterns of trade are developed.

In regard to climate change the first 100 years will be crucial. The extreme alternatives are either to stop anthropogenic emissions now and intensively restore plant life, or wait for the atmosphere's temperature to rise to a level at which all of the energy-gain from solar radiation could, once more, be radiated back into space. Since one is not feasible and the other is unacceptable, we can only hope that by completing some of the first part we shall be able reduce the effects of the second.

The rate of global warming is accelerating due to the build-up of atmospheric **GHG**, and since plant life is in decline there is little or no chance of reversing the trend until several years after all harmful activities have been stopped. The signs are not good:

Energy usage is increasing and we continue to use more hydrocarbon fuel.

The estimated time needed to develop a clean economy is doubling every 10 years.

There are only a few decades remaining in which any action to reduce the warming process might prove to be successful.

PART THREE

Figure 14

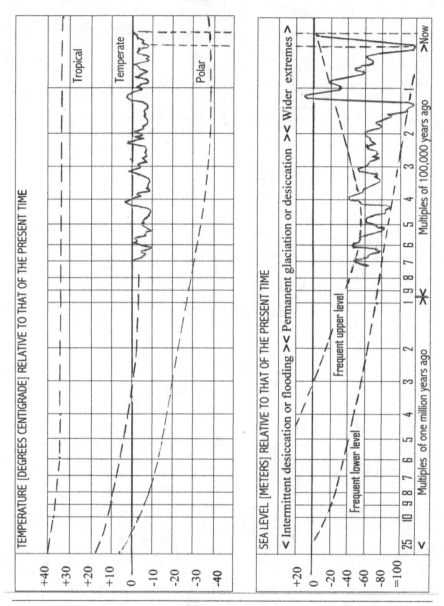

Temperatures and Sea Levels in the Neogene

PART THREE

RATES OF CHANGE
Figure 15

Foot Note	Greenhouse-effect Gas [GHG]	Global Warming Potential [GWP]	HALF LIFE YEARS	CONCENTRATIONS AT SEA LEVEL. [ppbv]	
				IN YEAR 1800	IN YEAR 2000
[1]	Carbon dioxide	1	50-200	288000	360000
[2]	Methane	21 to 23	8 to 12	750	1750
[3]	Nitrous oxide	296 to 310	130	160	315
[4]	SF_6	22200	3200	0.005	0.005
[5]	C_2F_6	11900	10000	0.003	0.003
[6]	CFC 11	4600	45	0.0	0.255
[6]	CFC 12	10600	100	0.0	0.545
[6]	CFC 113	6000	85	0.0	0.080
[6]	HCFC 22	1700	12	0.0	0.158
[6]	HFC 23	12000	260	0.0	0.014

Anthropogenic sources:
[1] Combustion of hydrocarbon fuels. Production of lime, cement, steel, aluminum, hydrogen, ammonia and fertilizers; agriculture, photosynthesis and metabolism.
[2] Agriculture, animal waste, metabolism, municipal waste, landfill, some industries.
[3] Production of nitric acid and Nylon; perfluorocarbons [PFCs], agriculture, fertilizers, legume-cropping, animal waste.
[4] Magnesium production.
[5] Production of aluminum and semi- conductors.
[6] Various halocarbons produced as solvents, refrigerants, aerosol propellants and foam expanders.

_____

GWP is the Global Warming Potential of a gas compared to that of an equal mass of carbon dioxide. Approximately 90 % of the GWP of a gas is the result of increased atmospheric water vapor as the air warms.
HALF LIFE is the time in years for half of the mass of a GHG to be removed from the atmosphere by prevailing conditions such as rates of dispersal, sorption, and biotic activity.

The Global Warming Potentials of GHGs

PART THREE

RATES OF CHANGE

Figure 16

TIME PERIOD BEGINNING Years ago	MEAN TEMP. CHANGE Deg.C	PEAK RATE OF TEMP. CHANGE (Deg.C)/Y	ENVIRONMENTAL CHARACTERISTICS
20,000	+3.8	+2.0 in 300Y	The beginning of the most recent great thaw: Sea level rose 120m, coasts were inundated, lowlands flooded, population followed the receding ice sheets.
13,000	- 2.4	-2.0 in 200Y	The 'Younger Dryas': Laurentide melt-water disrupted the Atlantic Current. Northern settlements were abandoned as another ice age returned. Loss of surface water run-off caused the Mediterranean Basin to dry.
10,500	+2.5	+2.0 in 150Y	Warming resumed and the sea rose 10m higher than its present level. The Mediterranean and Black Sea basins were re-flooded. Farming practices spread into Asia.
9,200	-1.0	-0.5 in 250Y	North American ice sheet collapsed. The Atlantic current slowed, causing a 'mini' ice age, while population continued to radiate from the Mediterranean zone. The fragile Saharan ecology became permanently stressed.
8,000	+1.8	+0.5 in 250Y	The Holocene temperature peak: Less cloud cover and a loss of vegetation.
5,500	-1.8	-0.5 in 250Y	Cooling cycle: Rise and fall of the Akkadian Empire in Mesopotamia. Severe droughts in the Middle East. The Sahara continued a cycle of desiccation. The Nile flow was reduced but continued to support Egypt.
3,000	+1.5	+0.5 in 200Y	Warming cycle: A longer growing season as the Mediterranean climate extended into Central Europe. Viticulture was widespread.
600	-0.8	-0.7 in 50Y	The Little Ice Age: Central Europe lost its sub-tropical climate. Iceland was inaccessible due to sea ice.
30 (1980 to 2010)	+0.7	+0.7 in 30Y	The Industrial Revolution began in 1770, accelerating a 13,000 year warming trend by burning more coal which increased air pollution. Today the warming is faster as GHGs from all fuels accumulate in the atmosphere.

Rates of Change in 20,000 Years

PART THREE

RATES OF CHANGE

Figure 17

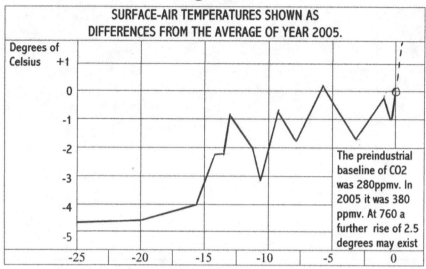

SURFACE-AIR TEMPERATURES SHOWN AS DIFFERENCES FROM THE AVERAGE OF YEAR 2005.

The preindustrial baseline of CO2 was 280ppmv. In 2005 it was 380 ppmv. At 760 a further rise of 2.5 degrees may exist

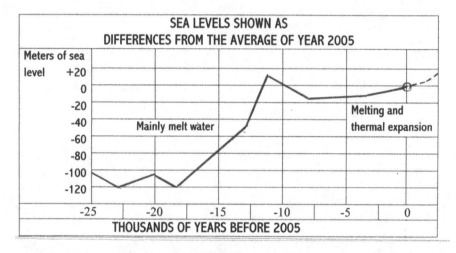

SEA LEVELS SHOWN AS DIFFERENCES FROM THE AVERAGE OF YEAR 2005

Mainly melt water

Melting and thermal expansion

THOUSANDS OF YEARS BEFORE 2005

Temperatures and Sea Levels 25KYA to the Present Era

PART THREE

Figure 18

Foot notes	Fuel Type	Heat energy GJ.t⁻¹	Carbon % by weight	Carbon (*) t.GJ⁻¹
(1)	Dry wood	11 approx.	90 approx.	0.082 avg
(2)	Methane	55	75	0.0137
	Ethanol	24.8	46	0.0185
	Butanol	29.2	57	0.0195
	Bio-diesel	37.8	87	0.0230
(1)	Coal	15-30	75	0.033 avg
(1] (3)	Crude oil	43	88	0.0205
(4)	Kerosene, Jet fuels	42.8	85	0.020
	Petro-diesel	43.3	87	0.020
	Gasoline	44.4	85	0.0191
(1) (5)	Natural gas, [mostly methane]	53	75	0.0142

(*) If combustion is complete, multiply carbon (t.GJ⁻¹) x 3.67 to convert to carbon dioxide (t.GJ⁻¹)

Heat energy is the amount of energy delivered to the user. The energy needed to produce and deliver it, can be subtracted to obtain the Net Energy Balance (NEB). The difference may be significant, especially in fuels produced in small quantities, e.g. bio-fuels.

(1) Natural fuels vary in heat-energy and carbon content.
(2) Bio-fuels are still in development. These data are for the compounds used at present.
(3) The recovery of usable fractions from crude oil can be up to 98 % of the amount processed, depending on quality of the crude and the need for specific distillates.
(4) Jet fuel energy and carbon content are practically the same as those of kerosene.
(5) Natural Gas heat energy is usually given as 36 GJ per 1000 m^3.

Hydrocarbon Fuels

PART THREE

RATES OF CHANGE

Figure 19

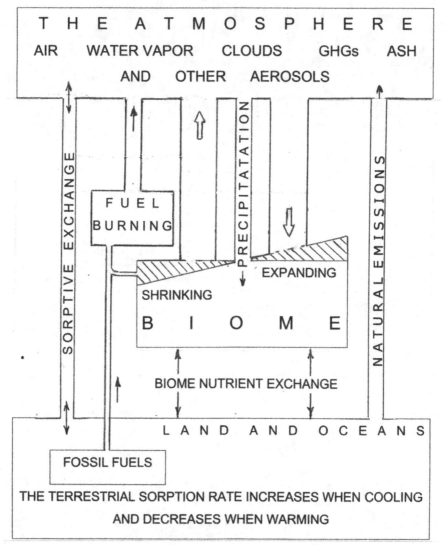

The Earth/Atmosphere Carbon Cycle

PART THREE

RATES OF CHANGE

Figure 20

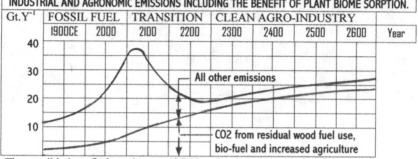

NET CO2 EMISSIONS IN AN IMPROVED AGRO-INDUSTRIAL WORLD. THE GRAPH SHOWS THE INDUSTRIAL AND AGRONOMIC EMISSIONS INCLUDING THE BENEFIT OF PLANT BIOME SORPTION.

The possible benefit from the use of CCS technology is not included.

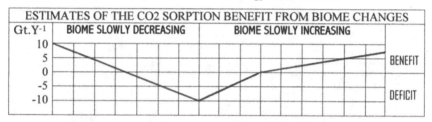

ESTIMATES OF THE CO2 SORPTION BENEFIT FROM BIOME CHANGES

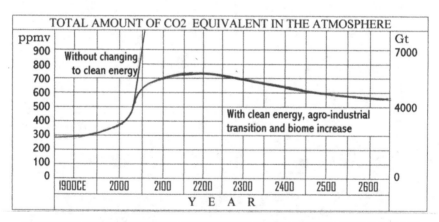

TOTAL AMOUNT OF CO2 EQUIVALENT IN THE ATMOSPHERE

CO2 Control in the Agro-Industrial Era

PART THREE

RATES OF CHANGE

Figure 21

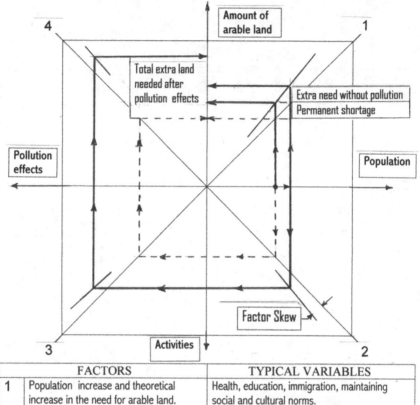

FACTORS		TYPICAL VARIABLES
1	Population increase and theoretical increase in the need for arable land.	Health, education, immigration, maintaining social and cultural norms.
2	Population increase and extra unpredictable human activities.	Poverty, expansion of trade, maintaining national independence and security.
3	Unpredictable human activities and greater incremental pollution.	Industry, construction, transportation, and maintaining over-all productivity.
4	Greater incremental pollution and its effect of reducing the profitability of national resources such as arable land.	World climate change, need to adapt agronomics to new conditions, loss of efficiency and lower crop yield per area.

Factors in The Agro-Industrial Transition

PART THREE

RATES OF CHANGE

Figure 22

REGIONS	ARABLE AND POTENTIALLY ARABLE LAND IN EACH REGION M Km²	AREAS CURRENTLY IN USE M Km²	CURRENTLY IN USE SHOWN AS PERCENTAGES OF REGIONAL TOTALS
East Asia and Pacific	7.8	4.78	61.3%
North Asia East of the Urals	3.0	1.75	58.3%
North America	4.8	2.33	48.5%
South and Central America	10.3	1.43	13.9%
Europe	3.8	2.14	56.3%
Sub-Saharan Africa	11.1	1.58	14.2%
North Africa and the Near East	0.7	0.7	100%
World (2007)	41.5	14.73	35.5%
(World. From an estimate dated 2001)	(50.2)	(14.0)	(27.9%)

Arable and Potentially-Arable Lands in 2007

PART THREE

RATES OF CHANGE

Figure 23

- Fresh-water wetlands have attracted human populations and sustained local economies for thousands of years. Their rate of disappearance (about half in the last 200 years) is now accelerating as they are drained to increase living space, or as new reservoirs are built to conserve water.
- Examples of lakes shrinking:
  CHAD
  ERIE
  MEAD
  NOP NUR
  ARAL SEA
- Examples of major rivers diverted or reduced:
  NILE
  INDUS
  YANGTZE
  MEKONG
  COLORADO
  RIO GRANDE
- Over 2.5 billion people are without domestic sanitation. Over one billion people are without clean drinking water.
- The depletion of aquifers results in the use of brackish water for crop irrigation which, lacking sufficient rain, permanently spoils the land.

Signs of Water Depletion

PART THREE

Figure 24

Year		2010	2060	2060	2200	2200
Global warming trend		now	A	B	A	B
World average ppmv of GHG as CO2 equivalent		380	650	1100	730	>2000
Subsequent av'ge temp. increase degrees Celsius		0	2	3	8	>10
ZONE above North Tropic	Popul'n billion	4.22	5.25	5.25	6.50	?
	LWF cu.km/Yr	29976	26730	39300	35000	?
	DWF cu.km/Yr	2870	6775	6775	6978	?
	DWF % of LWF	9.6	20.7	14.1	19.6	?
ZONE within North Tropic	Popul'n billion	1.79	2.37	2.37	3.80	?
	LWF cu.km/Yr	7632	6869	6687	5500	?
	DWF cu.km/Yr	880	2098	2098	2770<	?
	DWF % of LWF	11.5	24.7	25.4	49.3<	?
ZONE within South Tropic	Popul'n billion	0.61	0.91	0.91	1.80	?
	LWF cu.km/Yr	4032	3629	3527	3200	?
	DWF cu.km/Yr	440	1039	1039	1710<	?
	DWF % of LWF	10.9	23.4	24.2	52.7<	?
ZONE below South Tropic	Popul'n billion	0.18	0.24	0.24	0.40	?
	LWF cu.km/Yr	3360	3360	3192	2800	?
	DWF cu.km/Yr	190	449	449	622	?
	DWF % of LWF	5.7	7.3	11.5	21.9	?

LWF: Land water flow, is the estimated rainfall minus total evaporation.
DWF: Diverted water flow, is the estimated amount needed by the zone's population, borrowed from the LWF, used, then returned to the LWF. The DWFs needed in 2060 and 2200 are in proportion to estimated per capita usage in 2010, plus an amount for new farm development and solving present food and water deficiencies. Years labeled 'A' and 'B' follow different trends of CO2 concentration, from which their several DWF needs were determined. Those of 2060 are judged difficult to achieve, but possible in all zones with new water supply infrastructure; --of 2200A (marked <), very difficult in the tropics; --of 2200B, incalculable and dangerous in any zone.

Zonal Water Distribution from 2010 to 2200

PART THREE

PART FOUR

AN OVERVIEW

PART FOUR

PART FOUR

4.1 Perspective

Global warming is an Earth-process, increased at this time by what we continue to do rather than by any failure of remedies, though at our present rate of expansion it is unlikely that any remedy would suffice. The problem could be eased by slowing human activity and physically replacing all marketed combustion energy with clean energy, i.e. without waiting for fossil-fuelled power stations to be decommissioned due to their age. However the world is not even close to a reduction of combustion energy; in fact it is heading in the opposite direction. The amount of pollution emitted in the last 200 years is still raising the average air temperature, and even if all energy supplies were clean now, the trend already in progress would continue for decades; every day, increases the effort needed to correct it.

This is not the first time civilization has gathered sufficient strength to harm itself and the rest of the World, but in the last 50 years human activity tripled and the resultant **anthropogenic effect** showed us the extreme sensitivity of Nature. If we had lived at a slower pace there would be less need for concern.

Now we are endangered along with all other species, and it seems experience is telling us that we should have limited our expansion, and made humanity's presence more attuned to the World as it was before.

PART FOUR

Civilization has made its mark on Earth and there is nothing it needs to prove, except that it is capable of self control. Much of its worthwhile accomplishments existed long before the world was so crowded and before function defeated aesthetics. Few mementos of the modern era are likely to be worth preserving though museums have been built to house them. As for history, a **UNESCO Convention** to review world heritage sites was held in **1972.** Today UNESCO lists about 900 cultural and natural sites worth protecting, about 200 of which were added in the last 10 years.

Our generation should be remembered too, for choosing a better way to protect its descendants. But the timid steps taken thus far have achieved only temporary benefits. For example, city smog was reduced when catalytic converters were added to vehicle exhausts. It was relatively easy to do but the benefit was localized, and on its own, not enough to produce a significant reduction of **GHG** globally. The same happened in the 1940's when electricity began to replace open fires for domestic heating, and natural gas replaced coal gas. After these improvements smog was reduced but pollution returned in the post-war economic recovery.

As to our ability to limit global warming and climate change, we haven't gathered enough data to accurately predict the timeline of necessary effort. But there have been enough climate-related disasters to show that the cost of doing nothing will be substantial, probably as much as 10 or even 20% of **world GDP** annually.

Our first major step has to be the stabilization of human activity, the second, to fully compensate for it, as measured by the natural development of a greener **biome**. The third will be the ongoing development of a civilization that prefers to be an unobtrusive part of

Nature. Influential people from the time of Aristotle to the present day have tried to promote this idea but were always defeated by a popular desire to achieve 'greater things' and so we did. But we also lost the peace we desired in the world and in our lives.

4.2 Meetings and Reports

There has been no lack of official concern:

The **UN** held its first **Conference on the Human Environment** in **Stockholm (June 1972)** and commissioned a study using 152 consultants from 58 countries. Its report's title was **Declaration of the United Nations Conference on The Human Environment**. A summary was authored by economist **Barbara Ward (Jackson)** and scientist **René Dubos**. Its title was '**Only One Earth**' and with remarkable foresight it described most of today's conditions.

In **1983** the **UN** called for **'a global agenda for change'** to be provided by a World Commission on Environment and Development (**WCED**), and in **1987** it released its report '**From One Earth to One World**', or '**Our Common Future**'. This came to be known as the **Brundtland Report**, named after the WCED Chair, former Norwegian Prime Minister, Gro Brundtland.

At a **World Summit** in **Rio de Janeiro (1992)** the **WECD** published **Declaration on Environment and Development**, outlining twenty seven principles for international cooperation in a plan called **AGENDA 21**. The next scheduled meeting will be in **2012**. Actually, progress has been slow. **GHGs** are still raising the world's air temperature; for many years we have been causing more carbon dioxide to be emitted than the **biome** could remove.

125

PART FOUR

4.3 The Kyoto Protocol

The **KYOTO Conference** members signed a conditional agreement, to reduce **GHG** emission rates to 5.2% below the 1990 level by 2012. It was a slight ambition compared to what needed to be done. Several **Annex 1 countries**, meaning the main polluters except China and India, made their own individual commitments to emissions reduction goals. There were variations in the emission targets and in terms for achieving them, but it appears all will make an effort if not by 2012, then in the decades which follow. China and India, having large populations and growing economies, stated their intentions to reduce emissions if possible. The protocol itself allows nations some flexibility in how they can gain credits for their efforts. For example they may be purchased from another country which has met or passed its goals, or as a 'carbon-offset' from another country which has a lower cost of per-unit emissions reduction, or by investing in green projects like forest preservation, renewable energy and clean energy.

4.4 European Progress

Europeans might actually improve on the goals they set for themselves. In **2006**, economist **Sir Nicholas Stern** presented a report to the **UK** government showing the estimated cost of reducing climate change. Basically, the amount of carbon then in the atmosphere was about 40% higher than it had been in the pre-industrial era. Allowing the trend to continue would raise the cost of future mitigation twenty fold. The report called for successfully reaching targeted **GHG** reductions by 2050. Despite partisan controversy after publication, I believe the report was timely and correct in pointing out the urgent need to reduce global warming.

AN OVERVIEW

In 2009 the German **BMU** published **Renewable Energy Sources in Figures**, a detailed report on the status of world renewable energy. And in 2010 the **WWEA** held its **9ᵗʰ World Wind Energy Conference**. The UK, Germany, Spain and Norway are leaders in the amount of renewable clean energy produced. In regard to nuclear-power plants, France has been a leader in this field for at least 25 years but nuclear power safety has been brought back into question by the disaster at Fukushima, Japan beginning March 11, 2011. From the viewpoint of increasing global safety it makes other clean energy solutions even more important.

4.5 Clean Energy Progress

The present rate of increase is only one fifth that of total energy increase; the problem is that by starting to build it late, we have to deploy it even faster to catch-up. Making improvements at the same rate as the world's economy grows is simply not good enough.

It was shown in **Figure 23** (lower graph), that even if 90% of all applied energy came from clean sources the amount of carbon dioxide equivalent in the atmosphere would peak, at about twice today's total amount, and stay there for hundreds of years. As already mentioned, without clean energy it will go even higher and cause greater climate instability.

Fortunately the increased cost of fossil fuel and the downward trend of clean energy capital and operating costs, point in a helpful direction. Now the task is to find clean-energy solutions for each world region and for that information to be organized so that global management is possible.

There seems to be no international declaration of ways and means to quickly replace combustion energy with clean renewable energy. The **REN21 'Global Status Report. 2009 Update'** showed the total investment in renewable energy was 120 billion US dollars in 2008. It seems to me that could also have been expressed as merely 0.2% of **world GDP.** Therefore, on the basis that a guess at the requirement is better than none, I estimate that 1% of **world GDP** should be spent annually to stop using fossil fuel, and to install clean energy generators. A commercial literature review of the wide range of capital costs indicates that that amount might replace 1% of the world's total fossil fuel usage by having an *installed capacity* of 4% of the world's total energy consumption. At this rate, each year would reduce **GHG** emissions further. There is also the probability that in 30 to 50 years about four times that capacity could be purchased for the same cost (measured in 2010 currency). Global funding to sell ahead and lower average costs now, would be an incentive to reduce the timescale for stopping 90% of carbon dioxide emissions.

In regard to the world budget, it could be argued that there are other equally humanitarian goals to be achieved. But this concern is not the fault of environmentalism. The problem is the large amount of currency spent on non-essentials, which increases pollution at the expense of neglecting humanitarian goals. If we go forward in the old reckless manner climate change will take a larger human toll, and if we fail to improve on our present efforts the remedial cost, as already mentioned, will be greater. This will be discussed later in the book.

4.6 Emissions Control

We should review what is known about the rate of emissions compared to population growth: From 1990 to 2007 emissions grew 1.8% per year when population growth was 1.008% per year. From 2005 to 2007 it was 2.5% per year, when population growth was 1.515% per year. Evidently there are optimists waiting to see if emissions can be reduced without disturbing their own agendas for growth, and hoping to eliminate all but the mildest of remedies. Here are the pros and cons:

1. The **Stern** report's recommendation to spend 1% or more of **world GDP** on emissions control seems to be the only one available and should be in the world's budget. Its other suggestions were to pre-clean fuel and use it more efficiently. This might reduce the emissions-to-energy ratio 20% in 10 to 20 years. Carbon capture and storage (**CCS**) should also be an important part of any national plan, though its future use may not be as extensive as was first hoped.

2. CCS uses chemical engineering to select reliable processes and devices from a menu already developed for other uses. Its applications include the reduction of emissions in the production of usable fossil fuels and their fractions, the reduction of emissions from the ultimate use of those fuels, chemical processing, cement production, agriculture, furnaces, incinerators and the capture of gases from biological decomposition. It has the potential to stop more than 85% of carbon dioxide emissions wherever it can be properly applied. However there is little chance of its use on the majority of emission sources. Small applications would be too costly, and road vehicles impractical. Perhaps new power plants and older large installations will use it.

3. Estimates for the future use of CCS are largely based on scaling-up the modest amount which now captures less than 0.1% of the world's annual carbon dioxide emissions. The largest working systems at this time are used for natural gas and methane production in Norway, Canada and Algeria.

4. About 70% of the total emissions from all stationary engines are from coal, so this application of **CCS** could equal all other uses in the next 10 to 20 years. It all depends on the size of penalties which might be levied for continuing to pollute.

5. Known coal reserves have the potential to release about 5500 Gt of carbon dioxide. Everything else being equal, it would be better to leave the coal in the ground, and the same can be said for other fossil fuels, but obviously the total emissions also depend on how soon combustion can be replaced by clean energy, until that happens coal will continue to be mined. World emissions might be reduced by supplying fossil fuels onward, to countries already using CCS, in exchange for 'turn-key' clean-energy systems. The priority of where to install them would be decided on the basis of guarantees from the recipients to discontinue fossil fuel usage, and also on suitability of locations, terms of delivery and service contracts. The guiding principle would be to obtain the fastest reduction of the world total of carbon dioxide emissions.

6. Obviously, replacing coal depends on economics, yet there seems to be no definitive trade-off in using coal-power cleaned by CCS, instead of accelerating the use of new clean-power generators. There are other concerns: (a) Will it ever be feasible to decommission coal-fuelled generators? (b) If we keep them on-line will we be able to store the carbon dioxide? (c) Is storage safe? Earth's potential carbon dioxide storage capacity has been estimated at 2,000 to 11,000 Gt., and the lower estimate may not be sufficient any sudden release would be hazardous. Therefore the development and supply of renewable clean energy, should proceed as quickly as possible.

7. As economies grow, there are more waste materials. The organic components decompose releasing methane and other gases, which should be burned or converted to immobilize their carbon. The carbon dioxide from methane combustion has only 4.5% of the GWP of the unburned gas. There are plans to reduce methane release if it can be done economically. In 2004 the **US EPA** announced that 13 countries had joined the USA in its '**Methane to Markets**' program, to capture the gas and use it for fuel. In **2010** it became the **Global Methane Initiative.**

Though they are important, the above-listed controls will not prevent the rate of global warming from reaching a dangerous level, even if started immediately.

4.7 Conclusion of Part Four

As mentioned previously, the world was once on a better course. The combined potential of hydraulic, wind, solar and geothermal energy, could have been used for everything, but we couldn't refuse the energy so easily available from fossil fuels, and that accelerated **cultural evolution** toward technology and industrial output. Our lives now depend on the use of combustible fuels and we cannot be optimistic about an early turn-around. Instead we should face the situation as a reality:

- It is too late to prevent global warming; our only chance is to reduce its peak intensity.

131

PART FOUR

- Industrial expansion will continue because it is funded by the economy's influential service sector. The effect of this will be discussed later.

- The industrial nations might continue for a few more years declaring that further expansion is good, and developing nations will add to world pollution, making their own recovery even more difficult.

- In regard to clean energy, especially from wind farms and solar radiation, climate change will require us to build a variety of installations. Taking too much energy from only one source might have negative effects on regional ecological systems.

- The emerging timeline calls for the deployment of clean energy to be 10 or 20 times faster than in year 2010. Large arrays of wind turbines, photo-voltaic cells and solar-energy concentrators represent our main hope. There are understandable objections to their size, and also doubts concerning their future, but these are simply reflections of the large amounts of energy we intend to use when we should use less.

- Large offshore wind farms are working in Norway and the UK, but other nations may not have as much suitable coastline. Wave or tidal-turbines are in development to suit a wide range of conditions.

- Nuclear reactors are often called clean, but their fuel supply is limited and there is a disposal problem. It may be that new concerns about their ability to survive earthquakes and tsunamis will limit their future deployment.

- It seems that the absence of restrictions on total energy usage allows clean energy to be added to national grids without reducing fossil fuel.

AN OVERVIEW

- There is a benefit to be derived from conserving fossil fuel as a resource. One day it may hold more value for chemistry than for energy production.

Balancing the pros and cons, the world outlook includes an increase in uncontrolled global warming, climate change, and all of the accompanying problems. These will be part of our daily concern for the next 200 years. We cannot return completely to pre-industrial conditions, and because of our limited timeline whatever we attempt do now, whether it involves clean energy, bigger forests or **CCS**, can only be tried once and the sooner the better.

Eventually, when a state of maximum population is near, it should be possible to measure the cost of emissions in terms of human lives. At that stage, agriculture would undoubtedly be the last surviving energy-intensive occupation. The only remaining option would be to keep the human population and/or its activity, far below the saturation point.

Our previous ambitions caused us to think only in terms of expanding the economy to stay level with population growth. Civilization is indeed supported by billions of workers, who have to be paid, and the sheer volume of activity is a decisive factor in our future, but if we are going to add jobs they should have the effect of reducing pollution rather than increasing it.

PART FOUR

PART FIVE

PARITY AND INCOME EQUALITY

PART FIVE

PART FIVE

5.1 Perspective

For more than 3000 years, ambitious factions have tried to make gains at the expense of the masses. In the kindest of terms this was competition with no redeeming virtue. Our instincts through the ages haven't changed very much: For most people, competing is a lifelong occupation and is still the next best thing to winning. In self-defense they say that success is good for the economy, and if the alternative is failure there are few who would argue the point.

But this is about the failure of the system to provide equally. There is no economy, unless it provides for the needs of its populace, and today that economy is the entire World. Civilization grew too quickly for the benefits to be shared in an orderly fashion, and cultural evolution will retain this fault unless or until we choose more humanitarian criteria.

One of civilization's needs is for families to be self-supporting units but today there are millions who have little or no income. Concerns for this have been expressed in many studies, as for example in the **UK ODI's 'Briefing Paper 27 (October 2007)'**, and in **'Project Briefing No. 7 (January 2008)'**.

5.2 The Status Quo

At present, there are sufficient resources for all nations to share. The only unknowns are how much should be used and how much kept in reserve. That, and the world's population, determines whatever is possible. People could be equal without resources but that is not our intention; with them we can hope for a better life.

Value is only a unit of measure. Flow represents trade, and its *distribution* is what determines equality or the lack of it. The law of supply and demand applies to those with incomes rather than those with savings. Admittedly financial equality can be approached by taxing wealth and making the rich poor; but that wouldn't be a permanent solution. Equality requires perpetual effort, and civilization has not performed well in that regard.

Having found no other way to prevent poverty, the only way to approach the problem has been to deal with its worst symptoms.

Urbanization
The world's population centers have always been magnets for the unemployed and homeless. Sometimes these migrants have to build their own shelters, even without normal services like clean water and sanitation. The **UN ESA** report, **'World Urbanization Prospects. 2007 Revision'**, stated that 85% of governments were worried by this imbalance of distribution and 56% wished for significant de-urbanization. Without such intervention the number of town-dwellers in less-developed nations will have risen from 12% of world population in 1950 to approximately 58% in 2050, while rural populations will have shrunk from 56% to approximately 29%. In the same period of time, world population will have grown almost four fold.

An example of remedial planning can be found in India's **National Rural Employment Guarantee Act (NREGA)**. Introduced in **2005** it offers non-agricultural jobs in rural areas which helps to reduce poverty between growing seasons and supports public works programs. Stopping urbanization is difficult but similar programs could help build the rural economies of other nations, and by expanding agriculture it could slow the rate of global warming.

Exclusive or Unfair Trade

Exclusive or unfair trade denies the right of small companies and developing nations to have a fair share of trade volume. As early as the 1940's there was a movement in Europe to reduce the amount of exclusive trade. The result was the formation of a specialized federation with connections ranging from retail outlets, to standard labeling and the certification of 'goods and services'. The World Fair Trade Organization (**WFTO**) now serves as the administrative body, but the trade volume under its sponsorship is still very small. This might be due to lack of marketing funds or because most of the world believes international trade is sufficiently fair. But almost all trade is in bulk contracts which favor pre-existing relationships, thus putting developing countries at a disadvantage. There are also counterclaims that some developing countries achieve low manufacturing costs while failing to comply with Article 23 of the **Universal Declaration of Human Rights.**

The General Agreement on Tariffs and Trade (**GATT**) was established in 1949, and the first quadrennial UN Conference on Trade and Development (**UNCTAD**), was held in 1968. In 1995, GATT was transformed into the **WTO** and dedicated to the supervision and liberalization of trade. By 2000, trade agreements were failing due to continued dissatisfaction among small developing nations.

Subsidized Trade

Subsidization is used to keep one or more trades buoyant at the expense of others. Even in the absence of international trade it is used to support industries which are not profitable but deemed essential. In a sense, this is nationalization. The problem is in its use to gain market share by artificially reducing prices. If there is to be more equality in future, the international price of any item should reflect the real national cost of producing and delivering it to the point of sale. We may choose where we as individuals spend and buy, and this would be an advantage in an open market, but development should move the world toward narrower price ranges and truer costs. False pricing moves other nations toward protectionism.

Globalization

The difficulty often associated with globalization is that large corporations have used it to gain access to cheap labor markets and lower tax regimes, while having insufficient concern for the living conditions of local workers and their families. World trade has experienced setbacks due to this bad reputation. Selective globalization might be to blame as it distributes employment unevenly. Its humanitarian justification is that it already exists and workers depend on it, but the picture is changing: Countries are trying to develop their own resources to help supply the world and take full credit in international trade while repaying their loans. This can improve life for entire populations.

All nations have to find a place in an equitable distribution of world cash flow, which means being fairly paid in proportion to their share of world per-capita effort. In a totally equalized world, the best result would be to 'break even', growing or downsizing with the world average. At worst, a developing country would be unable to repay loans without an adjustment to the world balance. It is all about gaining equality through real-time economic development; growth is not necessarily a factor.

5.3 Achieving Parity

Given that we must use resources efficiently while protecting the natural world and advancing equality, there will be unavoidable complications in sharing human effort: For example, if we accept that trade prices should accurately reflect effort, who will decide the price of a commodity which is in short supply? Possibly its price should still be based on the effort invested to produce it but its distribution would be rationed according to need rather than going to the highest bidder.

Because we can always find such exceptions, our ideas of parity vary, especially in changing conditions. Whatever we may believe it to be, **parity** is not necessarily advanced by economic expansion, and might even be hindered by it.

- Its simplest definition is in finance, when the monetary units of two or more nations have equal purchasing power.

- For workers it is thought to occur when the average wage in two or more nations or employment categories, have equal purchasing power.

- For nations, it is thought to occur when their **GDP(PPP)** per capita amounts, are equal.

Such comparisons are always difficult, but we have to find a method which is generally accepted. The **GINI coefficient** which defines unevenness within sets of data, has the advantage of already being used in

141

economics, especially for studying income distribution. Problems only occur through our use of ambiguous data. For example, we can follow trends by studying 'incomes', but human development is more closely related to such things as the amount of disposable income, the number of dependants and the costs of living in their localities. Furthermore if educational costs, pensions and health care are human rights, shouldn't they be classed as tax-deductible essentials and excluded from the list of items to be bought with disposable income? It can be seen that simple averages are questionable, especially if we wish to make international comparisons.

It is important to use definitions that will endure and are transferable across international boundaries. My choice is to define parity in a way that includes exceptions without specifying all of them. If there are large families living on one income, nations should make allowances in tax and social benefits. Wage comparisons are acceptable if we understand the contingencies, yet the accuracy of measurement in a bad situation is less important than in a better one. Also, there will be fewer exceptions and greater accuracy when disparity is reduced.

One day, nations will know their conditions in relation to the world data set, and with some assistance will learn how best to develop parity within their borders and internationally. In the following example four stages of parity are suggested:

Stage (A). Local Parity means that the wages for specified types of employment have the same discretional purchasing power everywhere *within that locality*, provided that minimum wages and social benefits assure an adequate lifestyle in that locality.

PARITY AND INCOME EQUALITY

Stage (B). National Parity can only exist after Stage (A) has been accomplished and when the wages for the specified types of employment provide for equal discretional purchasing power *in all of that country's localities and regions, both rural and urban.*

Stage (C). Economic Community Parity could only exist after a number of nations achieve Stage (B) and form a community, organized such that the wages for the specified types of employment enable workers to have the same discretional purchasing power *everywhere within that community.* The GDP of that economic community would then be the sum of the **GDP (PPP)** amounts of its member-nations.

Stage (D). World Parity could only exist after all economic communities, nations, and other collectives achieve Stages (B) and (C), and when wages for the specified types of employment enable workers to have the same discretional purchasing power everywhere. World GDP would then be the sum of the GDP (PPP) amounts of its economic communities, independent member nations and other collectives.

The goal will be to reach a condition which completes Stage (C) and makes a real attempt to approach (D).

Note: Since parity will be affected by regional climatic change, it seems likely that the constraints of multi-regional consortiums such as ASEAN, MERCOSUR, NAFTA, and others, will have to be flexible.

Wage parity is only half of the goal. The other half is to consider the **Financial Standards, (FS),** of living, for households or families relying on specified types of employment. Of course individuals within them might earn less or more than the group averages, but might also have less or more expenses; there will always be inequalities within disciplines.

It is hoped that the national averages of household **GDP(PPP)** in various **FS** groups, will be compared in order to show the extent of such inequalities. It is also hoped that a world median of per-capita GDP(PPP), containing less-extreme financial standards, will begin to appear as nations arrange their economies to suit free and fair trade.

The old concepts, of 'fully developed' and 'developing' countries might then be replaced, or supplemented, by the concept of segments of the world's population having similar **FS** categories. In the following example, three categories are suggested. They ignore national boundaries, and treat all people as world citizens at various FS levels.

FS Category 1. The fraction of world population whose average income, expressed as per capita share of World GDP(PPP), is significantly below the world average.

FS Category 2. The fraction of world population whose average per capita income is close to the world average.

FS Category 3. The fraction of world population whose average per capita income is sufficiently above average to survive economic downsizing without undue hardship.

In the transition toward income equality, FS Category 1 would shrink, but, as a small amount of poverty is unavoidable, it would still exist. The same would apply to FS Category 3 in regard to a small amount of excessive income.

Reducing the number of workers in relatively lower and higher FS Categories increases the population in FS Category 2. This represents progress toward a new global economy. Financial analysts are anticipating how much money such a transition might add to the world's economic growth. It seems the World Bank and some developed countries have already taken this into account, but the downside is that such hopes for economic expansion are contrary to the need to reduce industrial activity, waste and pollution. Instead, the additional cash flow should be assigned to achieve humanitarian goals.

Other observers have said that parity would increase a capitalistic world-middle-class, which again raises the question of choice. The class feared by economists is one which consumes too many resources, demands too much profit and breeds its own population. It seems to exist now but it may not in future. Big-business influence should be diminished as small enterprises increase in number.

The migration of workers will also grow, and this represents a significant opportunity to mitigate the effects of climate change. Within this plan, parity would give two hundred million people per year a chance to work where they are needed, increase their annual incomes and alleviate shortages in their homelands.

Rates of change in this scenario are only provisional, though clearly there are time constraints forcing its implementation. For reasons already given, we need to combine the convergence of **FS** with global efforts to reduce emissions and grow more food. An extended timeline would not bother the leading industrial nations, but developing countries will want to see faster progress.

As a compromise, it might be possible to concentrate on urgent clean-energy initiatives first and phase-in regional agrarian transitions as soon as is possible. The greatest effort will be required now and in the next few decades, gradually easing as new infrastructure is completed. Later, as the world population approaches its maximum, cultural evolution will follow the pattern of whatever has proved to be successful, always with the proviso that it is accompanied by reasonable political stability.

All of the above variables make it difficult to estimate a time scale for income equality. As a guideline, **Figure 25** shows how the transition might proceed while all other human development projects are in-work. The monetary amounts are in year 2005 dollars.

Note: The population percentages are global, but they have to constructed from local and national amounts and be in accordance with parity Stages A, B, C and D, otherwise it would be mathematically possible for most of the poor to live in one country even when the figures shows improved global parity. That is the main difference between old methods of calculation and the new FS method.

The budget estimates allow **world GDP** to rise from 61 to 260 trillion (2005) dollars. Its rate of increase would be 0.51% annually during the first 200 years. The time scale for the remainder could be condensed, but if the plan were to take all of the time shown in the Figures, the subsequent increase would average 0.077% per year.

- The cash value of world GDP(PPP) per-capita increases fastest in the first 200 years, rising 37% from about 9500 to 13000 dollars.

- In **FS Category 1** the GDP(PPP) per-capita remains constant at about 3500 dollars, but its population drops from 66 to 10% of the world's total in the first 200 years and remains at that level.

- In **FS Category 2** the **GDP(PPP) per-capita** remains constant at about 9500 dollars, but its population rises from 16 to 70% of the world's total in the first 200 years, falling slowly thereafter to 17%.

- In **FS Category 3** the GDP(PPP) per-capita falls slightly from 28,500 to 27,800 dollars in the first 200 years with almost no change in its population. It then falls steadily to about 16,800 with its population rising from 19% of the world total to 73%.

Figure 26 shows the same plan from a different viewpoint. The ordinates have been changed to show percentages of world population in each FS Category.

Since Categories 1 and 3 approach parity from opposite ends of the scale, the change will require time to develop after the essential foundation is in place. A critical point in the transition might occur between years 2050 and 2100, when progress can be seen and measured. If industrial growth cannot be sufficiently restricted, continued global warming will shorten the time available for problem-solving and increase the cost of recovery.

5.4 Conclusion of Part Five

A scenario for promoting income equality has been described. As already mentioned, it is only one part of the total scene and its time span depends on the achievement of other global initiatives.

Our approach to equality doesn't avoid the issue of parity within nations as long as the data are taken from those which have, themselves, reached the standards previously mentioned. That is an integral part of the scenario.

147

PART FIVE

With this condition, an approximation of the 'coefficient of non-uniformity', loosely based on the **GINI coefficient,** can be calculated for various stages of progress. Two sets of figures are needed for each calculation. These are: World percentages of population in each **FS Category**, obtainable from **Figure 25** (lower table), and World percentages of **GDP(PPP)** in each FS Category, obtainable from **Figure 26** (lower table).

Within this scenario the approximate *global* coefficients of non-uniformity are then: 0.409 in year 2010, 0.215 in 2350 and 0.111 in 2750. Recently published GINI coefficients for *national* incomes are in the range 0.65 to 0.25. North American countries are midway between the extremes, and Scandinavians have the most equal distribution at 0.25. *(Sometimes a GINI INDEX is quoted, which is simply the coefficient multiplied by 100).*

The amount of progress is not immediately apparent from these numbers and is best viewed as follows:

The reduction of the global coefficient from 0.409 to 0.111 removes 73% of today's world income disparity.

The final coefficient of 0.111 is 225% better than the best found today in Scandinavia.

PARITY AND INCOME EQUALITY

Figure 25

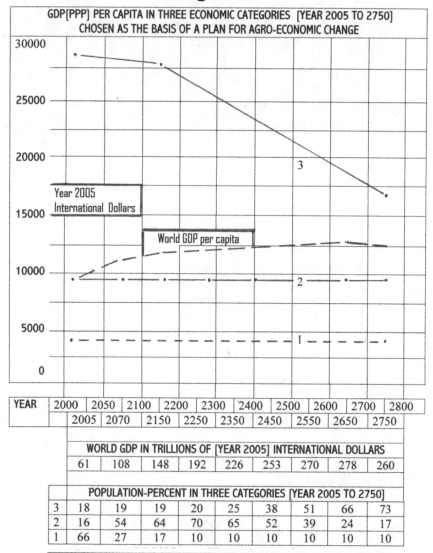

GDP[PPP] PER CAPITA IN THREE ECONOMIC CATEGORIES [YEAR 2005 TO 2750]
CHOSEN AS THE BASIS OF A PLAN FOR AGRO-ECONOMIC CHANGE

YEAR	2000	2050	2100	2200	2300	2400	2500	2600	2700	2800
	2005	2070	2150	2250	2350	2450	2550	2650	2750	

WORLD GDP IN TRILLIONS OF [YEAR 2005] INTERNATIONAL DOLLARS								
61	108	148	192	226	253	270	278	260

	POPULATION-PERCENT IN THREE CATEGORIES [YEAR 2005 TO 2750]								
3	18	19	19	20	25	38	51	66	73
2	16	54	64	70	65	52	39	24	17
1	66	27	17	10	10	10	10	10	10

Transition to Parity

149

PART FIVE

PARITY AND INCOME EQUALITY

Figure 26

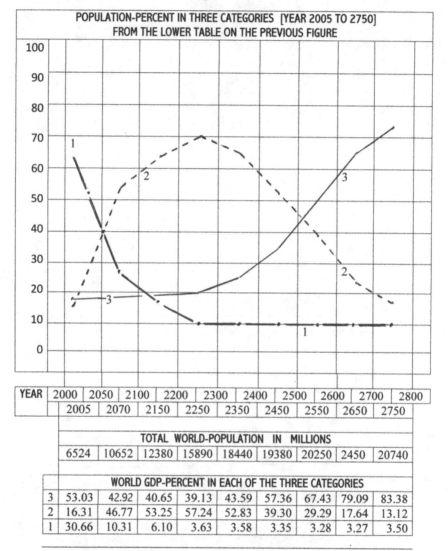

POPULATION-PERCENT IN THREE CATEGORIES [YEAR 2005 TO 2750] FROM THE LOWER TABLE ON THE PREVIOUS FIGURE									

YEAR	2000	2050	2100	2200	2300	2400	2500	2600	2700	2800
	2005	2070	2150	2250	2350	2450	2550	2650	2750	

TOTAL WORLD-POPULATION IN MILLIONS									
6524	10652	12380	15890	18440	19380	20250	2450	20740	

	WORLD GDP-PERCENT IN EACH OF THE THREE CATEGORIES								
3	53.03	42.92	40.65	39.13	43.59	57.36	67.43	79.09	83.38
2	16.31	46.77	53.25	57.24	52.83	39.30	29.29	17.64	13.12
1	30.66	10.31	6.10	3.63	3.58	3.35	3.28	3.27	3.50

Population in Three GDP(PPP) Categories

PART FIVE

PART SIX

HUMAN DEVELOPMENT

PART SIX

PART SIX

6.1 Perspective

The term **'human development'** is used to describe what we think should be done to further human well-being. The following definition might serve our purposes: We want sustained peace in a world of sufficient food, health, and equality, without a loss of cultural diversity. It's fair to say that this has faltered in the turmoil of economic growth, which takes most of our energy while more important steps are postponed. If properly managed with the combined efforts of a united world, human development could provide enough employment for all nations *and* reduce global warming by stopping pollution. The main headings, costs and timelines have already been presented as estimates.

To recap, the three main essentials are:

Restoration of the natural world. For safety reasons, it is important for the **biome** to recover as quickly as possible. The actions needed to mitigate the effects of pollution and global warming should be designed so that the amount of carbon dioxide equivalent in the atmosphere stays below the values shown in **Figure 20** (lower graph).

Improvement of nutrition and health standards. The amount of agricultural land should gradually increase to supply enough food for the populations shown in **Figure 12**.

155

PART SIX

Parity and income equality among nations. To ensure that long-term global tasks are supported by the appropriate regional effort, the benefits should be fairly distributed as shown in **Figures 25** and **26.**

6.2 Costs

Extra Administration and Service expenditures could start at 1% of **world GDP** annually, increasing to 2% per year by 2020 and then reducing to about 0.5% per year after 2050.

Figure 27 is a budget for the first eighty years shown in the form of a world capital-allocations table, listing the totals of all previously mentioned costs including Administration and Services. The purchasing power of the US dollar in 2005 is used and the amounts are given as percentages of world GDP. To simplify the presentation, the table shows annual averages in ten-year reporting periods. The highest rates of capital spending occur at the beginning, but systems maintenance would be an ongoing expense.

To improve long-term manageability there should be a limit to the number of regional initiatives active in each main topic at any one time. For example, each would be designed to last fifty years and by introducing them at five year intervals, no more than ten would be active at any time. The starting date, year 2000, has already passed but it is assumed that some of the easier steps have already been taken. Existing shortfalls in spending the appropriate amounts, will show how far the world lags behind what it should have done by now.

The total plan calls for spending 4.5% of world GDP in each of the first 20 years, and 3.8% in each of the following 30 years, subsequently falling to 2.3, 2.3, and 2.1% per year in ten year steps. The necessary 'starter' funding can be taken from other activities that should in any event be restricted. The following list shows a few of these, with estimates of their costs as percentages of **world GDP** in year 2010:

> 7 to 8% is spent on information technology,
> 4% on arts and entertainment,
> 3.5% on national armaments,
> 3% on new automobiles,
> 3% on crime reduction and the costs of prosecution.

These add up to about 21% of the world's GDP, and to fund new initiatives we need to recover just over one fifth (4.5%). This could be obtained by reducing the above-listed expenses in a number of ways:

- Spending on new automobiles could be halved by simplifying designs and using cars longer before trading them in.

- The cost of combustion energy will rise. Using clean energy will help to balance the budget.

- The total economic cost is partly compensated by a reduction of the effort now lost on unnecessary goods and services.

- The cost of humanitarian relief will be partly compensated by a reduction of the hidden penalty of global warming.

- On a per capita basis, resource consumption will be reduced, and less transportation will be needed.

- There should be less road-building and lower resurfacing costs if fewer private cars are used.

- Slowing industrial growth would allow capital to be redirected to global initiative funds.

6.3 Global Planning

There are many activities that waste resources, labor and energy. And there are others that support financial inflation, population growth and pollution. Stopping this deterioration could keep us occupied for centuries and probably will, but it won't happen simply because we wish it were so. There has to be a global plan supported by national goals.

The global effort might be helped by international advisers working out of a new **'Biome Development Council'** motivating and coordinating regional planning. After review and approval by local consultants, national governments would keep their assigned projects moving forward to a time schedule. World support would be encouraged by a view of better times ahead, for workers and their families. For most populations the great attraction would be more purchasing power, but it should be accomplished with 'greener' economic guidelines, the opposite of industrial consumerism.

In any event, the first task is to limit global warming. It seems the existing structure of world powers may not be optimum for this purpose as its priorities are influenced too strongly by other than humanitarian goals. Also, too few nations with too much influence, is the precursor of massive human error. Yet without cooperation there might only be stagnation; conferences with thousands of attendees would practically guarantee that nothing is done.

There are many ways in which democracy is layered and distributed. For example, debates reflect the interests of people, and the critics of change. One branch of national government could be dedicated to planning a humanitarian future, another to controlling the direction of research, and others limiting expansion and resource consumption. Perhaps the most urgent need at this time, is a branch coordinating the world's efforts to restore and protect the environment. At present this advice is acknowledged but then erased by immediate political and economic necessity.

As to our need to preserve cultural diversity, threatened as it is by the amalgamation of international efforts and purposes, if there are going to be factions, let them represent their people in a **World Cultural Democracy**. This will not be attained by calls for a new liberal policy or even by the idea of a global economy which is responsible only to its workers. Our interest is the condition of all people, and people can only reside in one culture not in the entire world. If their adopted nation is multicultural, as many are, then *that* is their new culture and each should contribute the comfort and security it offers following the principles of human rights. Clearly, powerful organizations representing hidden agendas can interfere in such a system. Educating the masses on how to separate their future needs from impossible dreams, would be a prerequisite to a global democracy. Developing countries, more than any others, need dreams, but too often they're influenced by powers which prefer the status quo.

Taking this further, it is worth considering whether or not at some future date, a **World Federation** might be established for the purpose of world planning. The concept is very old, but nation-states have always feared the loss of sovereignty and at this juncture rightly so. If confederation occurs at all, it will be after the world has shown it to be practical by gradually adopting its principles. If this fails, perhaps the closest approach to it will be further declarations of global purposes without the democratic support necessary to achieve them efficiently.

159

PART SIX

Whatever road is followed we should hope for a better future but without endless expansion. As it now stands, half of the global populace causes most of the environmental harm, and receives the greatest humanitarian benefits, while the other half tries to share the benefits by causing similar amounts of harm. The old belief in the need for economic growth is so entrenched as to make this spiral of deterioration virtually unstoppable. The old idea of unlimited human ability has to be scaled down; today we are witnessing the increase of many problems, while having insufficient time and skill to solve them. The future should include remedies and goals which are achievable without waiting on scientists and inventors to find compromises. We already have enough technical knowledge. Our task is to improve human conditions using conventional wisdom.

A stated goal, of the **UN** is to give all people a fair opportunity to develop in ways they value. Of course this statement hides an important caveat: We now understand that lifestyle development must stay within a framework of time-related limitations, otherwise it evolves randomly, hurting us and the World that keeps us alive. The rate of human development is a case in point. It has to catch up with humanity's increased power: not so slowly that we lose hope, nor so quickly that other problems increase.

To reiterate the logic: For safety reasons, the first priority will be to limit environmental harm. This need not delay human development. In fact, with so much work to be done, there will be opportunities in all countries for people to apply their skills. Some may need to learn new ones; we should ensure that their work helps to save the world, not add to its demise.

HUMAN DEVELOPMENT

Here are some propositions concerning the future, which could be debated:

1. Cultural diversity will be recognized as the ultimate need to ensure humanity's safety.

2. A world-federation will define long-term collective needs and apportion tasks to national governments. The world's management structure will be built from national representatives as today, but their tasks will have a more direct bearing on global well-being. National departments concerned with environmental improvement, will have access to world planning experts.

3. There will be more research into social comparisons between nation states, not for the purpose of measuring their fitness to compete but in terms of what needs to be done to attain optimal conditions for world unity. There will also be a possibility to give the world's populace more freedom to find work wherever there is employment.

4. It will be accepted that population control is impossible in a world of gross inequality. Greater parity will make it possible for all people to be taxed according to income alone, regardless of family size. Social benefits for large families will take the place of a graduated tax structure.

5. Within 100 years the economy will be well on its way toward primarily agrarian activities rather than primarily industrial ones. To this end, the economic purpose will resolve itself into finding useful work for all people while reducing emissions. The most likely solution will be a lateral shift away from industrial manufacturing and toward a greater content of human labor in the value of traded goods.

6. Within 150 years the world's sum total of employment will be equated to the average of human development. The overriding concern of each government will be to obtain employment which is compatible with and sufficient for, its country's humanitarian needs.

161

PART SIX

6.4 National Planning

The future can be imagined as it might be when the rate of global warming has been lowered and human development is more advanced. By that time nations should have a better understanding of their future. It will be useful to think in such terms very soon, and from there to imagine each country's per capita share in an improved global condition. Here are some propositions related to cultural evolution:

7. National success will be measured in terms of human development and all that that entails, and not as economic expansion.

8. In the 20th century, much effort was expended by the Service sector to find more efficient ways to use labor; in one way or another there were more people working on it than there were performing the actual tasks. In future there will be less management and more artisans, less routine and more individual skills.

9. 100 years from now, as the values of material resources increase, the need to increase human productivity will be diminished. Products will be made with more labor content and less machinery. This portends a future in which the emphasis will be on ways to live well with fewer resources.

10. The minimum disposable income after reaching maturity will be increased such that each person will be able to enjoy the benefits of the culture they help to develop. Workers without individual incomes, such as full-time home-makers and others unable to take paid employment, also those reaching the age of retirement, will receive all the other benefits of wage earners.

11. Developers of private care and health benefits which are beyond the reach of national average income earners, will do so at their own expense without subsidies, and be taxed at a higher rate.

162

12. The most useful areas for employment are likely to be in fresh-water distribution, sustainable mixed agriculture, stopping pollution, health-care cost-reduction, reduction of waste everywhere, the development of fair trade, and ensuring world parity in employment.

13. In 100 years, the excess of human resources will not be used to justify leaving the less-educated unemployed. The demand for completing standardized education prior to employment will be considered a relic of the industrial age. State-funded primary and secondary education will be a human right, regardless of age, ability or employment status.

14. The employment age related to state pension contributions will be from a minimum of 15 to a maximum of 55. Terms of employment will include opportunities for supplementary training or skill-related studies. The first 6 years will teach practical skills in a chosen discipline with the possibility to opt for part-time formal education. Afterwards it will be either at the discretion of the employer or self-funded.

15. Early pensions, for example at age 45, might be available at a reduced rate depending on the applicant's ability to be self-supporting, paying back any contribution deficiency by training others or by participating in voluntary work programs. Continuing studies might also be funded in this way.

6.5 Conclusion of Part Six

As national economies develop, they have to find a way to be compatible with global needs.

Planning is common to all people, but how they spend their time depends on where they live and on the culture which surrounds them. In the last 50 years it has been normal for countries in Asia and Europe to have more people than there are resources to keep them occupied. Those regions hold vast reserves of workers looking for employment abroad. China and India have this condition, yet if they were fully employed in today's economic scenario, resources would only be depleted that much faster.

As far as we have been able to tell, the world will continue to have centers of industry and regions for agriculture, but work distribution will depend on the conditions of populations, climates and transportation. Perhaps, the most difficult task of all will be achieving the best global balance from each nation's activity. There will have to be limitations on investments and on ownership entitlements otherwise commercial goals will continue to dominate human life.

The question is whether or not socio-economic revision can advance far enough and fast enough to effectively counterbalance the activity of nations set in motion 50 years ago. We may have the power to change ourselves, but whatever we choose has an effect on the World at large, and that may force us to adapt further.

If we could develop our cultures at a natural rate the adaptation to change would be eased and, perhaps for the first time ever, human activity could bring unqualified benefits. Until now we've raced to surpass previous limits, but we're gradually finding that Earth has its own limits. Our next test is to adjust, fairly and equitably, to this condition.

Fig 27

	2000	2010	2020	2030	2040	2050	2060	2070	2080
ALLOCATIONS TO LIMIT CLIMATE CHANGE, FUND AGRONOMICS AND ACHIEVE WORLD PPP. ITEMS ARE THE ANNUAL PERCENTAGES OF WORLD GDP AVERAGED IN EACH DECADE:									
Water supply	1.5	1.0	0.3	0.3	0.3	0.3	0.3	0.3	
Clean energy	1.0	1.0	0.5	0.5	0.5	0.5	0.5	0.3	
Admin,& Services	1.0	1.5	2.0	2.0	2.0	0.5	0.5	0.5	
Emission control	1.0	1.0	1.0	1.0	1.0	1.0	1.0	1.0	
Total annual spending as percentage of World GDP	4.5%	4.5%	3.8%	3.8%	3.8%	2.3%	2.3%	2.1%	

AVERAGED ANNUAL WORLD GDP. (TRILLIONS OF 2005 DOLLARS)								
	61.0	69.5	75.5	82.8	89.9	96.4	104	111.6

AVERAGED ANNUAL SPEND ON THE MAIN ITEMS (BILLIONS OF 2005 DOLLARS)								
Water supply	915	695	227	249	270	290	312	334
Clean energy	612	696	378	414	450	483	520	557
Admin & Services	612	1044	1570	1658	1800	483	520	557
Emissions control	611	695	755	829	900	964	1038	892
Average annual total spend.	2750	3130	2870	3150	3020	2220	2390	2340

GDP(PPP) Allocations to Year 2080

PART SIX

PART SEVEN

PART SEVEN

THE HUMAN CONDITION

PART SEVEN

PART SEVEN

7.1 Perspective

In a figurative sense the future is only time; a blank canvas waiting for a picture to emerge. Each generation looks ahead, as we ourselves do, but the problem in trying to imagine the future is that we can only use information from the past. Our situation is that we live within the laws of Nature on Earth and can learn more by studying recent history than by looking at the rest of the Universe.

Our daily concerns have always been survival, the social environment and gathering information. In terms of personal ambitions they can be redefined as health and safety, well-being, and organization. In all other ventures the structure of civilization is slowing down as it encounters new problems. We might believe that developments proceed according to a plan, but reversals begin at the first difficulty. This is one reason why history is so often rewritten; we prefer to measure progress by what has been achieved than by its cost.

In the modern era there have been many opportunities to make such compromises. For example, it's been said that the Industrial Revolution was caused by a need to perform arduous tasks with less effort. Another school of thought believed it was caused by the need to grow food while dedicating more time to social development. Yet another stated that mechanized farming released a workforce to take jobs in mining and manufacturing.

Today, it seems most of the world's activity is in the Administration and Service sector because it is 'the more efficient way'. Of course, these are only the points of view of industrial developers and have never applied to the majority of the world's workers. Such conditions might have developed anyway as a result of population increase; indeed they are still evolving.

Since perception is accountable for actions, I suppose the human condition can only be whatever we think it is, but cultures are not totally alike in this respect, and hopefully never will be. As it happens, success in one direction can defeat other endeavors; we can't always have both. For example the UN has been trying to increase human equality for decades, but now we've added the task of mitigating the effects of climate change. Raising the **HDI** may not have been the cause of global warming, but now it is a factor which complicates the task of reducing **GHG** emissions.

Because of our rate of activity, if not progress, the interactions between people and Nature are not at all what one finds in other species. The chains of events we cause are new and difficult to foresee. We are a radical element causing unnatural events.

7.2 The Search for Safety

There are occasions when evolution seems to reverse itself, returning to a previous format. It can happen either as a result of 'trial and error', or because of new unforeseen dangers.

It occurs in civilization too. Our safety is already compromised by technology which has been allowed to expand freely. I doubt if this is generally accepted to be a problem, and even less that over-achievement brings its own dangers. After all, we can imagine that tomorrow will be better than yesterday. But when viewed as part of world history, the ratio of progress to effort reached its peak about 5000 years ago. As mentioned in **Part 1**, and shown in **Figure 2**, the **innovation quotient** of the majority has been in decline ever since. There are more people than things to be accomplished, so the masses keep old skills to support, in pyramid fashion, the few who remain occupied with new projects.

It seems to me that the idea of people having traditional occupations is perfectly natural, but the new projects they are called to support are not. The danger of the intellectual pyramid is that it can direct or lead without reference to the well-being of its people. Collective progress might then be measured only from the leading edge, not from the base-line of the human condition.

The world's economy has developed in exactly that fashion, only the fact that we still live in separate nations and cultures, protects us from a world existing only to pursue a single dream. As mentioned several times, I believe diversity will remain as our main strength, yet now we are faced with global problems which require much greater cooperation than at any previous time.

In that event, humanity's most valued activities will be observing and comparing. Our ability to communicate and debate conclusions at a high level of complexity will be our most-used characteristic. The media will broadcast news, and social networking will help us keep our instinct for knowing the difference between right and wrong, true and false. That is not something that can be programmed into computers. The idea of virtual intelligence, or any other which detracts from our individuality should be set aside as counter-productive.

171

7.3 Finding New Goals

Beyond the scope of this study and far into the future, we will evolve according to cultural needs, but if we lose the desire to imagine, that will be the ultimate reversal, putting us on a path back to prehistoric times.

The signs can already be seen. In future, it will not be necessary for the masses to know how things work; not among manufacturers or marketers or sales departments, and certainly not among purchasers. As things become more complex they will have to be safer and more 'user friendly'. A significant amount of effort is already expended on this.

Eventually the choice will be between increased complexity at even greater cost or the division of tasks into smaller easier steps. I would imagine that our increased population will make the latter choice preferable. No doubt the technological age will run its course until the costs far outweigh the benefits, at which point a return to simpler work and a quieter life will be more appealing.

This would not necessarily be a reversal but it could be a step away from the world's present direction and we should be prepared to see its advantages. It could start now. The era of combustion energy should end, and the recovery of disappearing species should begin.

In regard to health care, there will be no reversal but the traditional way of allocating costs will be unable to meet the demand. The continued development of pharmaceutical products will require new ways of controlling their use, providing physicians with the information they need. The accumulated knowledge of drug interactions, nutrients, allergies and contra-indications, and the trend toward gene therapy, all point to a new branch of medical practice and the computerized storage of each patient's history.

The trend of mixing technology and science with daily life will continue, and the main divisions of science will merge in a way that enables the consequences of one to be measured by others. Scientific languages, which developed in each discipline to save discussion time, will gradually be unified, enabling greater freedom of debate between workers.

In future, grade school education will include the holistic appreciation of our place in Nature. At the present time we're trained to survive in the world as it is, without reference to its fitness for life. The demands of commerce direct our choice of careers, often before we are ready to decide for ourselves. There are few possibilities and fewer alternatives, and that deprives us of the right to look further and see if errors in the present system can be corrected.

7.4 Trends in Organization

Thus far, the future of organization was described vaguely as a **World Federation**. Its form might be an extension of what the **United Nations** was intended to be, but more effective in managing human development. In **2010, UNPA** in the **Campaign for the Establishment of a United Nations Parliamentary Assembly** published the **Declaration of Buenos Aires.** It requested that the UN proceed toward the organization of a parliament of member nations. As already mentioned, I think it would have to function informally first.

At the same time we were seeing the natural development of an answer to this need in the form of social networking. The only caveats are:

(a) It has to be universally available.

(b) It has to permit freedom of expression.

(c) It should debate principles but not individuals.

(d) New policies must be given a reasonable chance to develop before changes are seriously considered.

Just as there many levels of private communication, there should be more than one network: At least one for personal business, one for social or cultural topics, and one for governments.

As cultures focus more of their attention on global purposes, freedom of speech must be maintained to allow people to disagree with majority policies, even though they need to function within them. Similarly, private enterprises should be encouraged but controlled to align them with collective purposes; not necessarily without competing, but at least to keep them aligned with the best interests of the populace. Nation-states will also relate their own purposes to those of the global community, and will have the difficult task of balancing the needs of their cultures with those of the world.

It may be that the approaching crisis of human sustainability will convince nations to dedicate enough of their efforts to make a difference to the outcome. Studying the proceedings of trade blocs, alliances, and shared cultures would be helpful to understand how this might develop. This type of cooperation is the beginning of federalism and may serve as a pattern for the world.

7.5 Economic Revision

Seeing real issues from the viewpoint of economics, requires an appreciation of global causes and effects. Two thirds of **world GDP** is derived from the Service sector which is supposed to be non-polluting, yet it is not blameless in regard to global warming, and couldn't exist without a foundation of trade. If we say 'trade is the transfer of ownership', it follows that Service is also derived from that transfer. The alternative is to say that nothing is owned and everything is bartered, which is probably closer to the truth. Yet achieving it is causing an excessive amount of pollution which is robbing the world of the benefits of its efforts.

Redirecting the cash flow where it will do the most good will mean a return to some intermediate point in economic development. For the purpose of increasing the global **HDI** with a minimum of environmental harm, it seems that an *'agrarian and essential-industry'* mixture would be the good choice

Slowing and reversing any trend is difficult if it has significant momentum. Human activity is a prime example. Within the last 125 years, dedicated inventors introduced ideas which are still altering our lives. The list includes: the mass production of automobiles, civil aviation, automation, information technology and digital social networking. These and other innovations, with the activities they generate, represent over 90% of **world GDP.** The remainder, of less than 10%, is in agriculture. Together they pay for everything we use. In addition to food and shelter they pay for health-care systems, birth control, dietary supplements, antibiotics, faster emergency response, training, to enable us to live in such a world, and the inevitable luxuries we cherish but don't really need. Beyond that, our activity should at least improve the quality of life for people who have none.

There is no law that says how we should pay for such things, but they are not abstractions and all require work which has to be provided by the world's populace.

To improve the human condition we could just as easily obtain the funds by restoring the balance between industrial and agrarian values. Indeed, it seems that as food prices rise the transfer of value from industry to agriculture might already be in progress though its redistribution depends on securities-markets. Investors should believe that their financial health provides global benefits but it seems to me they are not overly concerned if the trickle-down effect goes no further than where it will be of greatest help to them. This will not, on its own, raise the world's **HDI**.

7.6 Peacekeeping

Even as the world's economy begins to recover from a series of reversals, military spending is increasing, and worse still it is using capital that could have gone to human development. Unanswered social grievances are part of the reason, international disparity and differences of belief account for the rest.

It is time for peace to be honorable in its own right. Each day the majority of the world's populace becomes more convinced that this is true, yet the human condition is such that very often, peace has to be enforced. Surveillance and verification is becoming important because, in the social fabric early attention is better than late repairs. Nations have to know about trends in vast populations; law and order depends, more than ever, on the abilities of their own forces.

When we examine today's global issues they all relate to the separation of people within and between cultures. But knowing and being able to do something about it are two entirely different things. Civilization's driving force comes from the world's most powerful committees who wish to retain their power. None are ready to say that it would be better to change than to continue as they are, and we have to admit that many show concern for the worsening situation and are willing to adapt. But the chances of pro-active cooperation have been dimmed by the same problems it needs to solve. The result is that each culture imagines it can only go forward with its original beliefs and ambitions intact.

This topic has been reviewed in great detail. Two reports are especially relevant; The UN's **Report of The Panel on United Nations Peace Operations. 2000 (The Brahimi Report),** and the UN report, **A New Partnership Agenda. Charting a New Horizon for UN Peacekeeping. 2009.**

7.7 Conclusion

Other species survive in balance with their surroundings, but humanity decided long ago that survival was not enough, and proceeded to devise its own environment. Yet in the final analysis the human condition is determined by the natural world, and within that limitation it also depends on its own beliefs and disbeliefs.

Our species was never able to live peacefully within these restrictions, though its factions might have found solace in rationalizing that they working for the greater good. At a higher level we know the difference between truth and wishful thinking. Too often goals are chosen for our personal advantage without regard for their collective usefulness or achievability. The result is that modern life can be amorally competitive and stressful.

Living with hopeful optimism stems from transposing history into the future in a way that suits our needs. There is some merit in believing it would be wiser to return to the best of what we have known, than to proceed into the unknown, yet globally there is no average past or future that all cultures might share, and in any event it is clear that the world is not ready to live by one set of rules. I prefer to believe that this is not merely an historical fact; it's how Nature caused things to evolve from the beginning. No culture working independently, will be able to point the best way ahead for the entire world, neither will any single government or trade alliance. Civilization has the unique responsibility of modeling international cooperation after the best it can find in Nature; that is what I've tried to explain in this book.

We need not look very far to find such a vision. Just beyond the fences of human habitation, there are views of awesome serenity and grandeur that give us a sense of what Nature really intended. It evolves as one entity while everything in it experiences change, surviving because of their interdependence. That is its eternal quality, a concept beyond morality and human intentions. It simply 'is', and we should learn from it.

Today's collective reality is that we are approaching Earth's limit to feed our population. In another world the limitation might have been insufficient water or oxygen, or the spread of persistent disease. It just happens that our limitation will be food, and apparently it calls for an **Agro-Industrial Revision** to help us adapt. That is not just a convenient supposition to fit this moment in time. Even if civilization had grown at a slower rate the result would have been the same, though we might have approached it with a greater margin of safety.

APPENDIX 1

REFERENCE TABLES

APPENDIX 1

REFERENCE TABLES

Figure 28

Large numbers:		Symbol:	Quantity prefix:
thousand	10^3	[K]	Kilo
million	10^6	[M]	Mega
billion	10^9	[G]	Giga
trillion	10^{12}	[T]	Tera
quadrillion	10^{15}	[Q]	Peta
quintillion	10^{18}	[E]	Exa

Linear: One meter =1m.   1000m =1Km.

Area: One hectare =1ha =0.01Km^2

Volume: 1m^3 =1000liters =6.2898 Barrels.

Weight: One pound (lb) = 454 grams (g).   1000Kg =1tonne (t).

Time: s=second.   h=hour. Y=year. YA=years ago.

BCE=before the Christian era.

Energy: One Joule =1J =1Kg.m^2/s^2

One calorie =1cal =4.19J.

One dietary calorie =1000cal.

British Thermal Unit [BTU].   1BTU =1055J.

Power: Watt [W].   1W=1J/s

Electricity: 1 BTU converted at 33% efficiency produces 0.00586 KWh

Large Numbers, Symbols, Conversion factors.

181

APPENDIX 1

REFERENCE TABLES

Figure 29

Speed of light.	300,000km.s^{-1} (approximately)
Solar system diameter (lens-shaped).	20,000,000,000km
The Sun's diameter.	1,392,000km
Mean distance from Sun to Earth.	145,000,000km
Solar radiation at the edge of atmosphere.	342W.m^{-2}
The Earth's mean diameter.	12740km
The Moon's diameter.	3475km
Mean distance from Earth to Moon.	384,000km

Age of the Earth.	4,550,000,000Y
Core material.	Iron
Mantle: Semi molten silica rock. Thickness:	2900km
Crustal rock thickness.	5 to 9km
Top soil: A thin fragile layer resulting from erosion, thermal, chemical and biological activity on rock formations and from organic growth and decay.	
Land area.	149,000,000km^2
Water-covered area.	361,000,000km^2

Earth's atmosphere, at present (approximately):

Nitrogen	78	% by volume
Oxygen	21	,,
Argon	0.925	,,
Remainder (various)	0.075	,,
Additional water vapor, average.	0.4	,,

Atmospheric strata; Troposphere	Below 10km altitude
Stratosphere (ozonosphere)	10 to 50km ,,
The ozone maximum is at:	20 to 30km ,,
Mesosphere	50 to 80km ,,
Thermosphere	80 to 650km ,, .
Exosphere	Above 650km ,,

Total atmospheric mass.	5.135x10^{18}kg
Total water vapor mass.	1.27x10^{16} kg

99.9999% of the atmospheric mass is below 100 km altitude.
75% of the atmospheric mass is below 10 km altitude.
50% of the atmospheric mass is below 3.2 km altitude.

Earth and its Atmosphere

APPENDIX 1

APPENDIX 2

VARIABLES IN POPULATION

APPENDIX 2

VARIABLES IN POPULATION

APPENDIX 2

The main variables in the population growth equation are shown in **Figure 30**.

M. The multiple of the starting population, after 'Y' years of change.

Y. The time, in years, to reach multiple 'M'.

F. The average child-bearing age per female.

N. The number of live births per female.

C. The percentage per year change of population.

All other factors are assumed to be constant during the time period '**Y**', though in reality they could change and alter 'C' independently, especially if the time span is over many generations. Some of the radicals are listed here.

- Ratio of males to females.

- Migration. Regions and cultures vary in their suitability for population growth.

- Unpredictable disasters.

- Ecological or genetic issues.

APPENDIX 2

- <u>Life expectancy at birth.</u> This factor includes a number of variables, such as national health standards, epidemic disease, health care, nutrition, sanitation, types of employment, and habitat (e.g. urban or rural). When life expectancy is changing it is difficult to isolate variables which are interactive.

- <u>Infant mortality.</u> National statistics vary from about 3 per 1000 in developed economies to about 150 per 1000 in the poorest of developing countries.

- <u>Access to sufficient food for a healthy life.</u> Today there is sufficient food but it is unevenly distributed. Thus far, efforts to improve distribution have not prevented malnutrition and starvation from occurring. The worst aspects of deprivation are found in overpopulated countries and in regions isolated by racial and/or political conflicts.

One of the concerns, stated at the **UN FAO Food Summit 1996,** was excessive population growth which could defeat agricultural output, first regionally and then globally. This may be exacerbated by rapid global warming and climate change which destabilizes food production. However these hazards are generally ignored, there is still no consensus on the need for population control, nor is there any on how it might be accomplished. Any legislation might affect the rights of individuals to follow their traditions or beliefs. Even so the world is aware of the arguments for and against voluntary control.

VARIABLES IN POPULATION

In the 1970's, the people of China were advised of their overpopulation problem, and the idea of smaller families was promoted to reduce poverty and hunger. It was also hoped to improve social services by reducing demand. By 1979 the policy of one child per family was implemented. China is now completing its first 25 years of 'one-child-policy', which is considered to be a qualified success. The average fertility has fallen from about five births per woman to less than two.

Critics of the one-child-rule, say there was no need for it, because a reduction was already occurring in response to national awareness of the problem.

APPENDIX 2

Figure 30

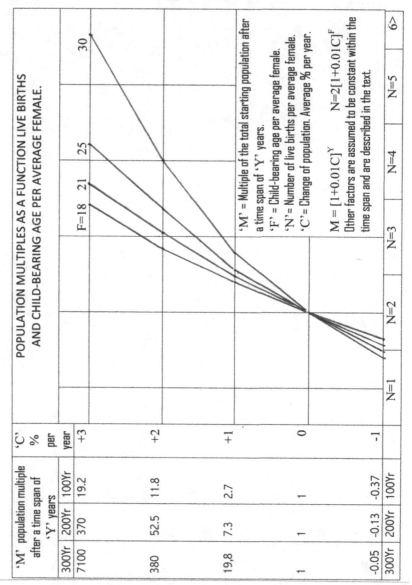

POPULATION MULTIPLES AS A FUNCTION LIVE BIRTHS AND CHILD-BEARING AGE PER AVERAGE FEMALE.

'M' = Multiple of the total starting population after a time span of 'Y' years.
'F' = Child-bearing age per average female.
'N' = Number of live births per average female.
'C' = Change of population. Average % per year.

$$M = [1+0.01C]^Y \qquad N=2[1+0.01C]^F$$

Other factors are assumed to be constant within the time span and are described in the text.

'M' population multiple after a time span of 'Y' years			'C' % per year
300Yr	200Yr	100Yr	
7100	370	19.2	+3
380	52.5	11.8	+2
19,8	7.3	2.7	+1
1	1	1	0
-0.05	-0.13	-0.37	-1
300Yr	200Yr	100Yr	

Population-change Equations

191

APPENDIX 2

GLOSSARY

Agro-Industrial Revision. An adjustment in the balance between agriculture and industrial output, necessitated by a shortage of food and excessive climate change.

Albedo. The ratio of reflected to incident solar radiation. Various coverings above and on Earth's surface, reflect from about 90% to 5%. They include, among others, snow, ice, water, atmospheric gases, aerosols, desert sand, forests, crops, soil, and asphalt. We have no way to measure Earth's effective albedo, but it is estimated to be 0.36. A higher value would reduce global warming, but our only hope of control is through the measurement and reduction of GHGs in the air. Reaching an improved condition quickly will require us to stop resisting it.

Anthropogenic effect. Pertaining to any change in the world, caused by human activity.

Biome. Earth's living things, the total of interdependent plants, animals and organisms.

Biome Development Council. A proposed advisory and implementation group concerned with the natural aspects of future life. Organized to gather input from geographic regions, make recommendations, and expedite subsequent initiatives by liaising with regional authorities.

Biosphere. Earth's living environment

Bolide. An asteroid or comet which has impacted Earth and released sufficient energy to produce a large crater.

Carbon dioxide equivalent. The hypothetical amount of carbon dioxide needed to equal the Global Warming Potential (GWP) of other pollutants in the atmosphere.

193

GLOSSARY

Clade. A taxonomic group containing species having a shared ancestry.

Climate change. A term commonly used to describe any observed redistribution or change, of weather.

CO2. Carbon dioxide. The main GHG released by the burning of bio-mass, bio-fuels and fossil fuels.

Collective. Pertaining to a group sharing habitat, culture, allegiance or beliefs.

Culture. The characteristics of species, nations and other collectives having biological and intellectual traits. Culture is applicable, with obvious limitations, to all living things. It may be found in the reactions of a collective to its living conditions, the way in which it functions rather than its physical presence, and therefore dependent on an ability to communicate internally. The cultures of sentient creatures may contain diverse functions but can be classified by shared instincts and beliefs.

GDP. Gross Domestic Product of a country, or group of countries, The market value of all goods and services produced within, regardless of ownership by external foreign corporations.

GHG. Greenhouse-effect gas. Any gas which, when released into the atmosphere, lowers Earth's albedo thereby raising its average surface temperature.

GINI coefficient. Named after statistician Corrado Gini. A measure of inequality within a population. The range is from zero (total equality) to 1.0 (total inequality). When multiplied by 100 it is the GINI INDEX. The coefficient for family incomes is generally in the range 0.2 to 0.7 but it doesn't necessarily relate to the average GDP per capita.

GLOSSARY

Global warming. Earth's average air temperature at ground level, is reported by authorities, such as NASA and UNFCCC. In natural history the temperature may rise or fall, as it is affected by Milankovitch cycles and changes in Earth's albedo. At present, the World is between ice ages, and is gradually warming. Pollution, from human activity, is accelerating the rates of warming, and climate change.

GNI. Gross National Income; GDP plus the income received from other countries.

GNP. Gross National Product. The market value of all goods and services supplied by residents of a country, including those from investments located outside of its borders.

GWP. Global Warming Potential. The ratio of the atmospheric warming effect of a gas, to that of the same volume of carbon dioxide.

HDI. Human Development Index. Obtained from data on life expectancy at birth, years of education and GNI(PPP) per capita.

Human Development. The unlimited process of the empowerment of people to live as they would want to live. In modern times this is associated with enabling less developed nations to rise closer to a world average.

Holocene. The most recent sub-division of the Quarternary period, which followed the Neogene. The present geological epoch.

Humanity. A collective noun for all of the world's people.

Innovation quotient. A measure of human inventiveness. The per capita rate of useful innovations is often higher in small collectives than in large ones, though the total rate of large collectives is greater. The lack of original ideas might cause us to think that, as a species, we have passed our intellectual peak, but that is only because we are pre-occupied with the needs of industry. It is the writer's belief that by taking a new direction, the general populace will show a new capacity for innovation.

Milankovitch cycles. Changes in the intensity and distribution of solar radiation reaching Earth's surface, caused by variations in the ellipticity of its orbits and in the tilt of its axis of rotation. The cycles so produced have their greatest effect at intervals of about 100,000 years and lesser ones at 50,000 and 25,000 year intervals. Named after Serbian astronomer Milutin Milankovitch.

Molecular clock. An imaginary device for comparing the rates of genetic mutation in species and the probable development of sub-species, based on data collected from previous evolution.

Natural. Existing, or having the possibility of being produced, without the use of machines. The opposite of manufactured.

Natural debt. That which is owed to the biome when a culture takes or destroys more than needed for survival as, for example, in the paving of farmland. A reduction of the World's capacity to support us.

Natural poverty. The result of allowing natural debt to continue. Repayment involves compensating the World wherever the loss occurs, without increasing the natural debt in other locations and habitats.

GLOSSARY

Nature. The collective appearance of laws which determine causes and effects everywhere at all times. We avoid saying the character of something is its 'nature'.

Neogene. The geologic period beginning 23.3MYA in which the continents reached their present positions and the global climate cooled to produce the most recent cycle of ice-ages; the time in which grasslands evolved and mammals of recognizably modern form appeared.

Philosophy. The rationalization of ideas existing in cultures. The extension of thought to include intangibles such as truths, moralities and ethics; built upon earlier schools of thought.

PPP. Purchasing Power Parity. A suffix for GNI, GDP etc., indicating that, to allow comparisons with data from other countries or regions, the amounts have been converted to a standard currency,

Sustainability. In today's language, 'sustainability' implies a desirable condition which can be maintained as long as is necessary. However, changes might be sustainable yet harmful; and there are some we might wish to stop. Therefore, we avoid using the word 'sustainable' without qualification of its purpose as in 'sustainable economy'.

Virtual water. The amount of water used to grow, and/or produce a tradable item and then export it to a water deprived nation. Generally applied to agricultural products.

Work. The expenditure of effort for any purpose over any length of time; not necessarily dependent on measurable success, remuneration or profit.

197

GLOSSARY

World. The proper noun is used for planet Earth as a living environment. The common case is used for a designated part, for example: 'the commercial world' or 'the inhabited world'.

World Cultural Democracy. An expression of the fact that each culture is a viable and necessary part of the **biome**, and that development should benefit not only its own populace, but those of all other cultures in an interdependent world.

World Federalism. A belief that international relations need to be coordinated for agreed practical purposes, and that a federal authority would develop a more cooperative future when guided by a World parliamentary system.

ACRONYMS: VARIOUS

ADB.   Asian Development Bank.
AFTA.   Asian Free Trade Area.
AOSIS.   Alliance of Small Island States
APEC.   Asia-Pacific Economic Forum.
ASEAN.   Association of Southeast Asian Nations.
BBC.   British Broadcasting Corporation. (UK).
BIS.   Bank for International Settlements.
BMU   Bundesministerium für Umwelt. (Germany)
    Federal Ministry for the Environment, Nature,
    Conservation and Nuclear Safety.
BP.   British Petroleum.
BRIC.   Brazil, Russia, India and China. (Alliance).
CAFTA.   Central American Free Trade Agreement.
CAIRNS.   The Cairns Group. (trade bloc).
CAP.   (See United Nations acronyms).
CARE.   Cooperative for Assistance and Relief.
             Everywhere. (US).
CBC.   Canadian Broadcasting Corporation
CCC.   Copenhagen Consensus Center.
CCEL.   Christian Classics Ethereal Library.
CCIR-NYC. Climate Change Information Resource of Columbia
             University.
CCS.   Carbon dioxide capture and storage.
CCSP   Climate Change Science Program (USA)
CDIAC.   Carbon dioxide Information Analysis Center.
             (USA). Also see ORNL.
CERF.   (See United Nations acronyms).
CETIM.   Centre Europe-Tiers Monde. (An NGO supporting
             human rights).
CFC.   Chlorofluorocarbon.

ACRONYMS: VARIOUS

CIESIN.	Center for International Earth Science Information Network at Columbia University.
CO2.	Carbon dioxide.
COMESA.	Common Market for Eastern and Southern Africa.
DOE.	Department of Energy. (USA).
EEC.	European Economic Community.
ECHA.	European Chemicals Agency.
ECOSOC.	(See United Nations acronyms).
EIA.	Energy Information Administration. (USA).
EPA.	Environmental Protection Agency. (USA).
E-PI.	Earth-Policy Institute.
EPI.	Economic Policy Institute.
EPS.	Economists for Peace and Security. (USA).
EU.	European Union.
FAO.	(See United Nations acronyms).
FDI.	Foreign direct investment.
GATT.	General Agreement on Tariffs and Trade.
GDP.	Gross domestic product.
GHG.	Greenhouse effect gas.
GILS.	Global Information Locator Service.
GISS.	Goddard Institute for Space Studies. (NASA).
GMO	Genetically modified organisms.
GMI.	Global Methane Initiative
GNI.	Gross national income.
GNP.	Gross national product.
GRI.	Geo-science Research Institute.
GWEC.	Global Wind Energy Council.
GWP.	Global warming potential (of 'greenhouse-effect gases').
HCFC.	Hydrochlorofluorocarbon.
HDI	(See United Nations acronyms).

ACRONYMS: VARIOUS

HFC.	Hydrofluorocarbon.
HIPC.	Heavily indebted poor country.
ICRC.	International Committee of the Red Cross.
IEA.	International Energy Agency.
IFPRI.	International Food Policy Research Institute.
IFRC.	International Federation of the Red Cross and Red Crescent Societies.
IMF.	(See United Nations acronyms).
IUCC.	(See United Nations acronyms).
LRAN.	Land Research Action Network.
MDG.	(See United Nations acronyms).
MDM.	Medicines du Monde. (Doctors of the World).
MERCOSUR.	The trade association of Argentina, Brazil Paraguay, and Uruguay.
MSF.	Medicines Sans Frontieres. (Doctors Without Borders).
NAFTA.	North American Free Trade Agreement.
NASA.	National Aeronautics and Space Administration. (USA).
NCPA.	National Center for Policy Analysis.
NGO.	Non-governmental organization.
NOAA.	National Oceanic and Atmospheric Administration. (USA).
NREGA.	National Rural Employment Guarantee Act. (India).
NREL.	National Renewable Energy Laboratory. (USA).
NSIDC.	National Snow and Ice Data Center. (USA).
OAU.	Organization of African Unity.
OCHA.	(See United Nations acronyms).
ODI.	Overseas Development Institute. (UK).

ACRONYMS: VARIOUS

OECD.   Organization for Economic Cooperation and
        Development.
OPEC.   Organization of Petroleum Exporting Countries.
ORNL.   Oak Ridge National Laboratory. (USA). Also see
        CDIAC.
OXFAM   Originally the Oxford Committee for Famine
        Relief, now OXFAM International.
PAGES. Past Global Changes. University of Bern.
        (Switzerland)
PERN.   Population-Environmental Research Network.
PBS.   Public Broadcasting System. (USA).
PFC.   Perflourocarbon.
PNAS.   Proceedings of the National Academy of Sciences.
PPP.   Purchasing power parity.
PRB.   Population Reference Bureau.
REACH   European regulatory database. Registration,
        Evaluation, Authorization and Restriction of
        Chemicals.
REN21   Renewable Energy Policy Network for the 21st
        Century.
SACN.   South American Community of Nations.
UIC.   Uranium Information Center Ltd.
UKMO.   United Kingdom Meteorological Office.
UNPA.   United Nations Parliamentary Assembly.
        (Campaign).
USDA.   United States Department of Agriculture.
USGS.   United States Geological Survey.
WEC.   World Energy Council.
WMO.   World Meteorological Office.
WRI.   World Resources Institute.
WWEA   World Wind Energy Association.

ACRONYMS: UNITED NATIONS

CERF.	Central Emergency Response Fund.
CFS.	(see FAO).
CSD.	Commission on Sustainable Development.
CTBTO.	Prep.Com. Nuclear Test Ban Treaty Organization.
DDA.	Department for Disarmament Affairs.
DESA.	Department of Economic and Social Affairs.
DGACM.	Department for General Assembly and Conference Management.
DM.	Department of Management.
DPA.	Department of Political Affairs.
DPI.	Department of Public Information.
DPKO.	Department of Peacekeeping Operations.
ECA.	Economic Commission for Africa.
ECE.	Economic Commission for Europe.
ECLAC.	Economic Commission for Latin America and the Caribbean.
ECOSOC.	Economic and Social Council.
ECPS.	Executive Committee on Peace and Security.
EISAS.	ECPS Information and Strategic Analysis Secretariat
ESA.	Department of Economic and Social Affairs.
ESCAP.	Economic and Social Commission for Asia and the Pacific.
ESCWA.	Economic and Social Commission for the Western Area.
FAO.	Food and Agricultural Organization.
FAO CFS.	FAO Committee for World Food Security
FCCC.	Framework Convention on Climate Change.
GRIDA.	Grid-Arendal. (United Nations Environmental Program information office).
HDI	Human Development Index.
HDR	Human Development Report.

ACRONYMS: UNITED NATIONS

IAEA.	International Atomic Energy Agency.
IBRD.	International Bank for Reconstruction and Development.
ICAO.	International Civil Aviation Organization.
ICC.	International Criminal Court.
ICSID.	International Center for Settlement of Investment Disputes.
IDA	International Development Association.
IFAD.	International Fund for Agricultural Development.
IFC.	International Finance Corporation.
IHO.	International Hydrological Association.
ILO.	International Labor Organization.
IMF.	International Monetary Fund.
IMO.	International Maritime Organization.
INSTRAW.	International Research and Training Institute for the Advancement of Women
IPCC.	International Panel on Climate Change
ITC.	International Trade Center.
ITU.	International Telecommunications Union.
IUCC.	Information Unit on Climate Change.
MDG.	Millennium Development Goals.
MIGA.	Multilateral Investment Guarantee Agency.
NGLS.	Non-Governmental Liaison Service.
OCHA.	Office for the Coordination of Humanitarian Affairs.
OHCHR.	Office of the UN High Commissioner for Human Rights.
OHRLLS.	Office of the High Representative for the Least Developed Countries, Landlocked Developing Countries, and Small Island Developing States.
OIOS.	Office of Internal Oversight Services.

ACRONYMS: UNITED NATIONS

OLA.	Office of Legal Affairs.
OPCW.	Organization for the Prohibition of Chemical Weapons.
OSG.	Office of the Secretary General.
PFII.	Permanent Forum on Indigenous Issues.
UNAIDS.	UN Program on HIV/AIDS.
UNCDF.	UN Capital Development Fund.
UNCTAD.	UN Conference on Trade and Development.
UNDCP.	UN Drug Control Program.
UNDP.	UN Development Program.
UNEP.	UN Environment Program.
UNESA.	UN Economic and Social Affairs
UNESCO.	UN Educational, Scientific and Cultural Org.
UNFCCC.	UN Framework Convention on Climate Change.
UNFPA.	UN Population Fund.
UN-HABITAT.	UN Human Settlements Program. (UNHSP).
UNHCR.	UN High Commissioner for Refugees.
UNICEF.	UN Children's Fund.
UNICRI.	UN Interregional Crime and Justice Research Institute.
UNIDIR.	UN Institute for Disarmament Research.
UNIDO.	UN Industrial Development Organization.
UNIFEM.	UN Development Fund For Women.
UNINSTRAW.	UN Int'l Research & Training Institute for the Advancement of Women.
UNITAR.	UN Institute for Training and Research.
UNODC.	UN Office on Drugs and Crime.
UNOG.	UN Office at Geneva.
UNON.	UN Office at Nairobi.
UNOPS.	UN Office for Project Services.
UNOV.	UN Office at Vienna.

ACRONYMS: UNITED NATIONS

UNRISD.	UN Research Institute for Social Development.
UNRWA.	UN Relief and Works Agency for Palestine Refugees in the Near East.
UNSECOORD.	Office of the UN Security Coordinator.
UNSSC.	UN System Staff College.
UNU.	UN University.
UNV.	UN Volunteers.
UPU.	Universal Postal Union.
WBG.	World Bank Group.
WCED.	World Commission on Environment and Development.
WEO.	World Economic Outlook (of the IMF).
WESS.	World Economic and Social Survey.
WFP.	World Food Program.
WHO.	World Health Organization.
WIPO.	World Intellectual Property Organization.
WMO.	World Meteorological Organization.
WTO.	(1)World Trade Organization.
WTO.	(2)World Tourism Organization.

RESEARCH: BOOKS AND REPORTS

Brock, J., Webb, J.W. 1968. A geography of mankind.

Camerer, C., Loewenstien, G. 2002. Behavioral Economics.

Canadian Hydropower Association and others. (2000). Hydropower and The World's Energy Future.

Carson, R. L., 1962. Silent Spring.

Cohen, J.E. 1995. How many people can the earth support?

Colborn, T., Dumanoski, D., Myers, J. 1996. Our stolen future.

Conroy, G.C. 1997. Reconstructing human origins.

Fagan, B. 2004. The long summer.

Flannerty, T. 2006-2007. We are the weather-makers.

Fuguoka, M. 1987. The road back to nature.

Gordon, A., Suzuki, D. 1990. It's a matter of survival.

Gore, A. 2006. An inconvenient truth.

Gowlett, J.A.J. 1992. Ascent to civilization.

Hansell, R.I.C., Hansell, T.M., Fenech, A. A New Market Instrument for Sustainable Economic and Environmental Development. Journal. Environmental Monitoring and Assessment. Volume 86, Numbers 1-2/ July, 2003.

Hopfenberg, R., Pimentel, D. 2001. Human population numbers as a function of food supply.

Houghton, R. 2004. Understanding the global carbon cycle. The Woods Hole Research Center.

Jackson. See Ward (Jackson), B.

Jacobs, J. 2000. The nature of economics.

Kelso, A.J., Trevathan, W.R. 1984. Physical anthropology.

Kinder, H., Hilgemann, W. 2003. The Penguin Atlas of World History.

RESEARCH: BOOKS AND REPORTS

Kohak, E. 1999. The Green Halo. A Bird's-Eye View of Ecological Ethics.

Kovach, R., McGuire, B. 2004. Guide to global hazards.

Kuhn, T.S. 1970. The structure of scientific revolutions.

Liberal-International. 1997. Oxford Manifesto. The liberal agenda for the 21st century.

Margulis, L. 1970. Origin of Eukaryotic Cells.

Schneider. S. 1997. Laboratory Earth.

Sen, Amartya K. 1999. Development as freedom.

Shutler, R. Jr., (editor). 1983. Early man in the new World.

Tachi Kiuchi. 1997. What I Learned in the Rainforest. Keynote address. World Future Society.

Taylor, A. Global finance. Past and present. 2004. Finance and Development. March 2004.

Thomas, H., Thomas, D.L. 1941. Living biographies of great philosophers.

Tibbs, H. 1991. Industrial ecology.

Triandis, H.C. 1995. Individualism and collectivism.

Ward (Jackson), B., Dubos, R. 1972. Only one earth.

Wells, S. 2002. The journey of man.

RESEARCH: WEBSITES

AFRICA. Africa strives to revitalize agriculture.
www.un.org/ecosocdev/geninfo/afrec/vol1no2/overview.htm

ADB. The 2007 issue of the Asian Economic Monitor.
www.adb.org/

BBC. The Neolithic Revolution.
www.bbc.co.uk/dna/h2g2/A2054675

BBC NEWS. Planet under pressure (a six part series).
http://news.bbc.uk/1/hi/in-
depth/sci_tech/2004/planet/default.stn

BBC NEWS. World Trade Blocs. 1999.
http://news.bbc.co.uk/hi/english/static/special_report/1999/11/
99/seattle_trade_talks/default.stm

BIS. Comparison of creditor and debtor on short-term external debt.
Paper No. 13. 2002. http://www.bis.org/publ/bispap13.htm

BIS. Financial Technology Congress. 2008. Keynote Address.
www.bis.org/speeches/sp081119.htm

BMU. Renewable Energy Resources in Figures.
www.bmu.de/files/english/renewable_energy/

BP. Statistical Review of World Energy. June 2010.
www.bp,com/,,,/bp.../statistical_energy_review.../2010./statistic
al_rewiew_of_world_energy_full_report_2010.pdf

BRIC. Countries Joint Communique. www.globalpolicy.org/

BWEA. British Wind Energy Association. (See UKWED)

CARBONIFY. Current carbon dioxide levels.
www,carbonify.com/carbon-dioxide-levels.htm

CBC. DOC ZONE. The Disappearing Male
www.cbc.ca/documentaries/doczone/2008/disappearingmale/ch
emicals.html

CCSP-USA. US Climate Change Science Program. Final Report
(2009). www.climatescience.gov/

RESEARCH: WEBSITES

CDIAC. www.cdiac.ornl.gov

CDIAC._Global fossil-fuel CO2 emissions.
http:// cdiac.ornl.gov/trends/emis/tre_glob.html

CETIM. Supporting human rights. www.cetim.ch/

CHINA. South-to-North water diversion project. Water
Technology. www.water-
technology.net/project_printable.asp?projectID=2658

CHINA. Strategy and mechanism study for promotion of circular
economy and cleaner production in China. Council for International
Cooperation on Environment and Development. 2003.
www.harbour.sfu.ca/dlam/Taskforce/circular.html

CHINA DAILY. 2004. Pilot project mapped for green economy.
www.chinadaily.com.ca/english/doc/2004-07/23/content-
350791.htm

CHINA. Diversion to relieve drought.
www.china.org.cn/english/21390.htm

CIA. World Fact Book. US.
www.cia.gov/cia/publications/factbook/index.html

CIESIN. Rosenzweig, C., Parry, M.L., Fischer, G., and Frohberg, K.
(1993) Climate change and world food supply.
www.ciesin.org/docs/004-046/004-046.html

CLUB OF MADRID. www.clubmadrid.org

CLUB DE PARIS. (Paris Club). International economics.
www.clubdeparis.org

CLUB OF ROME. www.clubofrome.org

COPENHAGEN CONCENCUS CENTER.
www.copenhagenconcensus.com

DOE. Renewable energy trends.
www.eia.doe.gov/cneaf/solar.renewables/page/trends/rew_links
.html

RESEARCH: WEBSITES

DOE. www.sustainable.doe.gov/articles/indecol.shtml Also on www.eere.energy.gov

DOOLEY. J. 2001. Carbon dioxide capture and geologic storage. Joint Global Change Research Institute. USA.

EARTH IMPACT DATA BASE. www.unb.ca/passc/ImpactDatabase

EARTH POLICY INSTITUTE. http://www.earth-policy.org

ECHA, What is REACH? (EC 1907/2006)
http://echa.europa.eu/doc/timeline_en.pdf
http://ec.europa.eu/environment/chemicals/reach/reach
intro.htm

EGYPT. Toshka Project. Desert reclamation. Planet Ark feature. www.planetark.org/dailynewsstory.cfm?newsid=9340

EGYPT. Toshka and East Oweinet. South Valley Development Project. Civil Engineer.
www.icivilengineer.com/Big_Project_Watch/South_Valley/

ELDIS DATABASE. Aid and Debt. Heavily Indebted Poor Countries Initiative. Status of Implementation. www.eldis.org/cf/search/disp/docdisplay.cfm?doc-DOC19824&resource-flaid

ENCYCLOPEDIA OF NATIONS. UNDP. UN. Development Program. Technical cooperation programs.
www.nationsencyclopedeia.com/United-Nations/Technical-Cooperation-Programs-Ev.

ENERGY DATA BASE LINKS. www.edte.org

ENERGY WATCH GROUP. www.energywatchgroup.org

EPA. (US). Clean energy. Air emissions. www.epa.gov/cleanrgy/emissions.htm

E-PI._Earth-Policy Institute. Brown, L. Eco-economy updates. www.earth-policy.org/Updates

RESEARCH: WEBSITES

EPI. Economic Policy Institute. Weller. C.E., Scott, R.E., Hersh, A.S. The unremarkable record of liberated trade. www.epinet.org/content.cfm/briefingpapers_ sept01inequality

EPS. Economists for Peace and Security. www.epsusa.org/

FALUDI, J. 2004. Concrete. A burning issue. http://Worldchanging.com/archives/001610.html

FAO. (see UN FAO)

GILS. Global Information Locator Service. www.gils.net/index.html

GISS. Surface Temperature Analysis.
htpp://data.giss.nasa.gov/gistemp/graphs/

GLOBAL FLOOD MAP.ORG
http:// globalfloodmap.org/

GLOBAL ISSUES. Poverty Facts and Stats. www.globalissues.org/TradeRelated/Facts.asp

GLOBAL POLICY FORUM. (A consultative NGO). Tables and charts on UN finance. www.globalpolicy.org/finance/tables/index.htm

GLOBAL WARMING. A closer look at the numbers. http://clearlight.com/-mhieb/WVFossils/greenhouse_data.html

GLOBAL WARMING. The main greenhouse gases. http://grida.no/climate/vital/05.htm

GMI. Global Methane Initiative.
www.globalmethane.org/

GREENPEACE INTERNATIONAL.
www.greenpeace.org/international

GREENPEACE INTERNATIONAL et al. Concentrating Solar Power. Outlook 2009.
www.greenpeace.org.international/.../concentrating-solar-power-2009/

RESEARCH: WEBSITES

GWEC. Global Wind Energy Outlook 2010
www.gwec.net/fileadmin/documents/.../GWEO%202010%final.pdf

HUMAN POPULATION THROUGH HISTORY.
http://desip.igc.org/populationmaps.html

HUMANITARIAN ORGANIZATIONS.
www.geneva.ch/Humanitarian.htm

ICRC. International Committee of the Red Cross. www.icri.org/

IFPRI. Von Braun, J. 2007. The World Food Situation.
www.ifpri.org/pubs/agm07/jvbagm2007.asp

IEA.International Energy Agency. www.iea.org

IEA.World Energy Outlook. 2009.
www.worldenergyoutlook.org/2009.asp

IEA.Trends in Photovoltaic Applications. Report.IEA-PVPS-T1-19:20
www.iea-pvps.org/

IMF. Dollar., Kraay. 2001. Trade Growth and Poverty.
www.imf.org/external/pubs/ft/fandd/2001/09/dollar.htm

IMF. Watkins, K. 2002. Making globalization work for the poor.
www.imf.org/external/pubs/ft/fandd/2003/03/watkins. htm

IMF. Strauss-Kahn, D. 2010. Human Development and Wealth Distribution.
www.imf.org/external/np/speeches/2010/110110.htm

INDIA. GOVERNMENT OF DELHI. Atmosphere. Compendium of statistics. Chapter Four.
http://delhigovt.nic.in/newdelhi/dept/economic/env/chapter4.htm

INDIA. GOVERNMENT OF INDIA. Agriculture.
http://pib.nic.in/archive/pprinti/milestones2002/milestones-02-agri.html Also on
http://pib.nic.in/feature/feyr98/fe0798/PIBF2107983.html

213

RESEARCH: WEBSITES

INDIA. GOVERNMENT OF INDIA. Population.
http://pub.nic.in/archive/lyr2000/rjul2000/r22072000.html

INDIA. GOVERNMENT OF INDIA. Ministry of Water Resources.
www.wrmin.nic.in/   also, Water. Press Information Bureau.
http://pib.nic.in/feature/feyr2002/fsep2002/f20092021.html

INDIA. NREGA. National Rural Employment Guarantee.
http://india.gov.in/sectors/rural/national_rural.php

INDIA. UNITED STATES CONSULATE, MUMBAI. Global poverty
results uneven. http://mumbai.usconsulate.gov/   and
www.hwashnews1665.html

INTERNATIONAL WATER RESOURCE ASSOCIATION.
http://wc.Worldwatercongress.org:5050/

IPCC. Reports available on,
www.ipcc.ch/
www.grida.no/climate/ipcc
www.gcrio.org/ipcc
http://unep.no/climate/ipcc
> Climate change. 1996, 2001.
> Synthesis report. 2001.
> Mitigation. 1990-2010.

IPCC. Fourth Assessment Report on Climate Change.
www.wmo.int/pages/partners/ipcc/index_en.html

IPCC. Special report on carbon dioxide capture and storage.
http://ipcc.ch/activity/srccs/index.htm

IPCC. Technologies, policies and measures for mitigating climate
change. www.gcrio,org/ipcc/techrepl/

IPCC. Special report on emissions scenarios. Chapter 3. Scenario
driving forces. 3.2 Population.
www.grida.no/climate/ipcc/emission/051.htm

IPCC. AR4 SYR Synthesis Report-1 Observed changes in climate
and their effects.
www.ipcc.ch/publications_and_data/ar4/syr/en/mainsl.html

RESEARCH: WEBSITES

KYOTO. UN, FCCC. Framework Convention on Climate Change.
http://unfccc.int/resource/docs/convkp/kpeng.html

LRAN. Land Research Action Network.
www.landaction.org/display.php?article=1

Montana Edu. Milankovitch cycles and glaciations.
www.homepage.montana.edu/-
geol445/hyperglac/time1/milankov.htm

NASA. GISS. Goddard Institute for Space Studies. Simulating the
1951-2050climate with an Atmospheric-Ocean Model.
www.giss.nasa.gov/research/briefs/sun_01/

**NASA. GISS. Goddard Institute for Space Studies. Surface
Temperature Analysis: Graphs.**
http://data.giss.nasa.gov/gistemp/graphs/

NASA. SPACELINK. http://usgs.gov

NATION MASTER. Geographic regions, economics.
http://nationmaster.com

NEO. Near Earth Objects. Information Center.
www.nearearthobjects.co.uk

NGLS Round-up report index.
www.unsystem.org/ngls/documents/publications.en/roundup/i
ndex.htm

NGLS Roundup report. (2008) UNCTAD XII.
http:// www.un-ngls.org/IMG/pdf/RU_133_4.pdf

NIH, National Institute of Health (US). www.nih.gov

NIH, Battle of the Biofuels, Manuel, J.
www.pubmedcentral.nih.gov/articlerender.fcgi?artid=1817686

NOAA, The NOAA annual greenhouse gas index (AGGI)
www.esrl.noaa.gov/gmd/aggi/

NOAA and others. File:Radiosonde satellite surface temperature.
png htpp:// en.wikipedia.org/wiki/File: Radiosonde_satellite
_surface_temperature.png

RESEARCH: WEBSITES

NOAA. Climate History: Exploring climate events and human history. www.ncdc.noaa.gov/paleo/ctl/cliihis.html

NOAA. The Mid-Holocene "Warm Period". www.ncdc.noaa.gov/paleo/globalwarming/holocene.html

NSIDC. National Snow and Ice Data Center. www.nsidc.colorado.edu/cryosphere

PNAS. Proc. Natl. Acad, Sci. USA. Vol95, pp. 14009-14014. November 1998. Applied Physical Sciences. Social Sciences. Hypsographic demography: The distribution of the human population by altitude.

OECD Agricultural Outlook 2005-2014. OECD/FAD2005 www.oecd.org/.../0.2340.fr_2649_201185_35015941-1-1-1-1.00.html

OECD. Organization for Economic Cooperation and Development. www.oecd.org

OUR PLANET.COM. Population and atmosphere. http://ourplanet.com/aaas/pages/atmos01.html

ORNL.Bioenergy. http://bioemergy.ornl.gov/papers/misc/energy_conv.html

PACIFIC NORTHWEST NATIONAL LABORATORY. 2006. Atmospheric Science and Global Change Division. Carbon dioxide capture and storage analyzed for role in addressing climate change. www.pnl.gov/atmospheric/highlights/2006052506.stm

PERN. Population-Environmental Research Network. www.populationenvironmentalresearch.org

PHYSICAL GEOGRAPHY. Solar radiation, atmospheric effects, temperature, (etc). http://www.physical_geography.net/fundamentals/7f.html

POPULATION REFERENCE BUREAU. www.PRB.org

POVERTY MAP. www.povertymap,net

RELIEF ORGANIZATIONS. www.reliefweb.int/

RESEARCH: WEBSITES

UKWED. UK Wind Energy Database . 'renewablesUK'.
www.bwea.com/ukwed/index.asp

**REN21. Renewable Energy Policy Network for the 21st Century.
Global Status Report. 2009** Update.
www.ren21.net/Portals/97/documents/GSR/RE_GSR_2009_Upda
te.pdf

STERN REPORT to the UK government. 2006.
www.combusem.com/STERN.HTM

TED. Case Studies. Case number 264. China. Three Gorges Dam
Project. www.american.edu/projects/mandala/
TED/THREEDAM.HTM

THINK TANK LISTINGS. www.mira.go.jp/iee

UK. Foreign & Commonwealth Office. Global Issues.
www.fco.gov.uk/en/global-issues/

UK ODI. Project Briefing No. 7
www.odi.org.uk/resources/download/399.pdf

UK ODI . Briefing Paper 27
www.odi.org.uk/resources/download/5.pdf

UN. The Millennium Development Goals Report 2010
http://mdgs.org/ened/mdg/Resources/Static/Products/Progress
2010/MDG_Report_

UN. Universal Declaration of Human Rights.
www.un.org/Overview/rights.html

**UN. A New Partnership Agenda. Charting a New Horizon for UN
Peacekeeping. 2009**
www.un.org/en/peacekeeping/documents/newhorizon.pdf

**UN. (Brahimi Report). Report of The Panel on United Nations
Peace Operations. 2000**
www.unrol.org/doc.aspx?n=brahimi+report+peacekeeping.pdf

RESEARCH: WEBSITES

UN. Declaration of the United Nations Conference on The Human Environment. 1972
www.unep.org/Documents.Multilingual/Default.Print.asp?docum
entid=97& article

UN. ECE. HIGH COMMISSION FOR EUROPE. www.unece.org

UN. ESA. Economic consequences of population ageing.
www.un.org/esa/policy/wess/wess2007files/chap4.pdf

UN. ESA. World migrant stock. The 2005 revision population base.
http://esa.un.org/migration/index.asp?panel=5

UN. ESA/CSD. Economic and Social Affairs/Commission on
Sustainable Development. www.un.org/esa/sustdev/csd

UN. ESA. World Urbanization Prospects. The 2007 Revision.
www.un.org/popin/data.   and.
www.un.org/esa/publications/wup2007/2007WUP_ExecSum_we
b.pdf

UN. FAO.
-   Conference. Rome. 2010.
-   The State of Food and Agriculture, 2009.
-   Land and Water Development Division. Aquastat. 2005.
-   Terrastat. 2003.
-   Toward 2015/2030. 2003.
-   World Food and Agriculture. NAS Colloquium, Rome 1998.
    Alexandros, N., Outlook.... World Food Summit, Food Crisis,
    Rome 2008.
-   Land tenure thesaurus.
-   Spatial Data and Information.
-   GIEWS. Global Information and Early Warning System on
    food.
-   Potential Impacts of Sea-Level Rise on Populations and
    Agriculture. 1998. www.fao.org/sd/eidirect/eire0045.htm
-   Twenty Ninth FAO Regional Conference for the Near East.
    NERC/08/INF/5. Annex 1. Climate change impacts.

218

RESEARCH: WEBSITES

Aquastat database.
www.fao.org/nr/water/aquastat/countries/inde/stm

Declaration of the World Summit on Food Security. 2009
www.fao.org/fileadmin/templates/wsfs/Summit/Docs/Final
_Declaration/WSFSO5

UN. General Assembly, September 18, 2000. Resolution 55/2 UN
Millennium Declaration.

UN. General Assembly, GA/10547. December 6, 2006. Arms trade
treaty, nuclear-weapons-free World, outer space arms race, etc.
www.un.org/News/Press/docs//2006/ga10547.doc.htm

UN. HCR. High Commissioner for Refugees.
www.unhcr.org/basics.html

UN. Millennium Project. http://unmillenniumproject.org/

UN. NGLS. Non Government Liaison Service. http://un-ngls.org/

UN. OCHA. Office for the Coordination of Humanitarian Affairs.
http://ochaonline.un.org/

UN. OCHA. Office for the Coordination of Humanitarian Affairs.
2002. Paying the ultimate price. Analysis of the deaths of
humanitarian workers (1997-2001).
www.reliefweb.int/symposium/PayingUltimatePrice 97-01.html

UN. ODC. Countering the global problem of corruption.
www.unodc.org/unodc/index.html

UN. ODC. Crime and development in Africa.
www.unodc.org/newsletter/en/200503/page005.html

UN. ODC. United Nations Crime Congress: 50 Years.
www.unodc.org/newsletter/en/200502/page002.html

UN. Peace and Security through Disarmament.
http://disarmament.un.org/

UN. Program on AIDS. www.unaids.org

UNCTAD. (Shafaeddin, M., Free Trade or Fair Trade?)
www,unctad.org/en/does/dp...153.pdf

219

RESEARCH: WEBSITES

UNDP. Human Development Report 2010. The Real Wealth of Nations.
www.hdr.undp.org/publications/hdr2010/en/HDR_2010_EN_
Complete.pdf

UNEP. Environmental program. www.unep.org

UNEP. World Conservation Monitoring Center.
www,unep-wcmc.org/latenews/index.cfm

UNEP. Vital Water Graphics; Water use and management.
www.unep.org/dewa/assessments/.../water/vitalwater/links.ht
m
UNEP. Grid Arenal. Vital graphics net.
www.vitalgraphics.net/climate2.cfm
www.grida.no   and on   www.povertymap.net

UNEP. Grid Arenal. World's surface water: precipitation,
evaporation and runoff.
http://maps.grida.no/go/graphic/world-s-surface-water-
precipitation-evaporation and runoff

UNESCO. United Nations Educational, Scientific and Cultural
Organization.
www.unesco.org

Convention Concerning the Protection of the World Cultural and Natural Heritage.
whc.unesco,org/archive/convention-en.pdf

Water use in the World. Present situation/future needs. .
www.unesco.org/science/waterday2000/water_use_in_the_Wo
r ld.htm

World water development report.
www.unesco.org/water/wwap/

World water resources and their use.
http://espejo.unesco.org.uy/

2003. International year of fresh water. Newsletter.
www.wateryear2003.org/en/ev.php-
URL._ID=1607&URL_DO=DO_TOPIC&URL_.

**Water. A shared responsibility. Second World Water
Development Report. 2006. (Virtual Water)**
www.unesco.org/water/wwap/wwdr/wwdr2/pdf/wwdr2_ch
_11.pdf

UN. IDO. Various documents connected with industrial
development on www.unido.org /doc24839 Renewable and rural
energy./doc51260 Poverty reduction through productive
activities./doc51262 Environment./doc51363 Energy and
environment.

UN. WFP. World Food Program.
www.wfp.org
www.un.org/issues/m-food.asp
www.thehungersite.com

UN. WHO._World Health Organization
www.who.int
www.who.int/publications
World Health Report 2008
MDG Goal 6, combat HIV/AIDS, malaria and other diseases
(2008)

UN. WMO. World_Meteorological Organization_
www.wmo.ch/wmo50/e/World
www.earthinstitute.columbia.edu

UN._World Bank Group_ Definitions and stats. index.
http://Worldbank.org/edstats/indicators.html
Also Documents and reports. www.Worldbank.org/servlet/

UN,_World Bank Group. Agricultural Biotechnology. Technical
paper no.133. 1991. World Bank, Agricultural and Rural
Development Department, Washington. D.C.

RESEARCH: WEBSITES

UN, World Bank Group. Byerlee, D., Diao, X., Jackson, C. Agricultural, rural development, and pro-poor growth. ARD Discussion Paper 21.

UN. World Bank Group. 2005. Mini atlas of millennium development goals. Building a better World.

UN. World Bank Group. (Data and statistics). http://web.Worldbank.org/WEBSITE/EXTERNAL/DATASTATISTICS

UN. World Bank Group. World Development Indicators. http://devdata.Worldbank.org/

UN. World Bank Group. Development Report 05. A report on world progress toward the achievement of the UN's Millennium Development Goals. http://devdata.Worldbank.org/wdi2005

UNIVERSITY OF BERN. SWITZERLAND. Past Global Changes. Research Project. www.pages.unibe.ch/science/research/newstructure.html

UNIVERSITY OF BRITISH COLUMBIA OKANAGAN. Solar radiation. Atmospheric effects. Temperature. www.physicalgeography.net/home.html. Search items 7f and 7m.

UNIVERSITY OF CALIFORNIA. Atlas of Global Inequality. http://ucatlas.ucsc,edu/

UNIVERSITY OF CALIFORNIA BERKELEY. Press release 2005/08/02. Faster carbon dioxide emissions will overwhelm capacity of land and oceans to absorb carbon. www.berkeley.edu/news/media/releases/2005/06/02_carbon.shtml

UNIVERSITY OF MICHIGAN Library database. Foreign and international economics. www.lib.umich.edu/govdocs/stecfor.html

UNIVERSITY OF WATERLOO, CANADA. Evolutionary economics and the counterfactual threat. 1999. www.cgl.uwaterloo.ca/-racowan/counter.html

RESEARCH: WEBSITES

UNPA. Campaign for the Establishment of a United Nations
Parliamentary Assembly International Meeting. 4 Oct. 2010.
(Declaration of Buenos Aires).
www.unpacampaign.org/documents/en/2010outcome.pdf

USDA. Economic Research Service. Gelhar, M., Coyle, W.
Global food consumption and impacts on trade patterns.

USDA, Estimating the Net Energy Balance of Corn Ethanol. 1995.
www.ethanol-gec.org/corn_eth.htm

USDOE. International Energy Outlook 2010
www.eia.doe.gov/oiaf/aeo

USEIA. Annual Energy Outlook 2010
www.eia.doe.gov/oiaf/forecasting.html

USGS. Geological Survey.   http://usgs.gov
    Sea level and climate. http://pubs.usgs.gov/fs/fs2-00/

WCED. Conference on Environment and Development. Agenda
21. 1992. Rio de Janeiro (Agenda 21)
www.un.org.geninfo/bp/enviro.html

WCED. The Brundtland Report. 1987
UN. Report of the World Commission on Environment and
Development. 1987. Also known as ': From one earth to one
world' and also by the title 'Our common future'.
www.un-documents.net/wced-ocf.htm

WEBEC. Economics database.
www.helsinki.fi/WebEc/webecc8d.html

WEC. World Energy Council. Survey of resources.
www.Worldenergy.org/wec-geis/publications/reports/ser/

WEC. The Twentieth World Energy Congress. Rome 2007.
www.rome2007.it/congress/Congress.asp

WOODS HOLE RESEARCH CENTER.
www.whrc.org/carbon/index.htm

WORLD POPULATION BUREAU. World population data sheets.
www.prb.org

RESEARCH: WEBSITES

WORLD WATER CONGRESS.
http://WC.Worldwatercongress.org

WRI. La Vina, A., Fransen, L., Kurauchi, Y., Faeth, P. Beyond the DOHA round and the agricultural subsidies debate.
www.wri.org

WTO. International trade statistics. Most frequently accessed tables.
www.wto.org/english/res_e/statis_e/its2005_e/its05_bysubject_e.htm

WWEA. World Wind Energy Report. 2009
www.wwindea.org

WWF. The Energy Report. 100% Renewable Energy by 2050.
http://assets.panda.org/downloads/101223_energy_report_final_print_2.pdf

INDEX

[Illustration page numbers are in bold print)

225

226

INDEX

[Illustration page numbers are in bold print)

INDEX
[Illustration page numbers are in bold print)

INDEX

[Illustration page numbers are in bold print)

INDEX

[Illustration page numbers are in bold print)

INDEX

[Illustration page numbers are in bold print)

INDEX

I

INITIATIVES, REGULATIONS AND REPORTS

INDEX

[Illustration page numbers are in bold print)

[Illustration page numbers are in bold print)

INDEX

[Illustration page numbers are in bold print)

NATURAL POVERTY. 67, 196
NATURE. 3, 42, 65, 124, 125, 169, 170, 178
   196:
   Molecular Clock. 37, 196
   Mutations. 37
NEOGENE. 73, 197, **99**

O

ORGANIZATION. 16 et seq.:
   Agricultural Revolution. 10
   Consortiums. Multi-regional:
      BRIC. 13
      ASEAN, MECOSUR, NAFTA. 143
   Democracy. 20, 48, 158 et seq., 174, 198
   Education. 45, 65, 142, 163, 173 et seq., 195
   EEC. European Economic Community. 14
   EU. European Union. 14
   European Renaissance. 11
   Federalism. 159, 161, 173, 198
   Harappa, Indus Valley. 10
   Health care. 87, 172
   Industrial Revolution. 14, 169, **103**
   Infrastructure. 17
   Intellectual pyramid. 171
   Olmec, Mexico. 10
   UN. United Nations. 17 et seq., 173, **25, 27, 29, 31:**
      IBRD, IMF, World Bank. 18
      NGO, NGLS. 19
   World. 158 et seq., 173, 176, 177
   Yangshao, China. 10

P

PHILOSOPHY. 39, 65, 197
POPULATION. (see HUMANITY)
PPP. 141, 148, 197, **149, 151, 165**
PURPOSE. 48

INDEX

[Illustration page numbers are in bold print)

QR

READING OR REFERENCE:

S

SCENARIOS:

INDEX

[Illustration page numbers are in bold print)

INDEX

[Illustration page numbers are in bold print)

X Y Z